Life-Cycle Assessment

Inventory Guidelines and Principles

B.W. Vigon, D.A. Tolle, B.W. Cornaby, and H.C. Latham
Battelle-Columbus
and
C.L. Harrison, T.L. Boguski, R.G. Hunt, and J.D. Sellers
Franklin Associates, Ltd.
and
U.S.E.P.A. Risk Reduction Engineering Laboratory

LEWIS PUBLISHERS
Boca Raton Ann Arbor London Tokyo

DISCLAIMER

The information in this document has been funded wholly by the United States Environmental Protection Agency (EPA) under Contract No. 68-C0-0003 to Battelle. It has been subjected to peer and administrative review, and it has been approved for publication as an EPA document. Mention of trade names or commercial products does not constitute endorsement or recommendation for use. Use of this methodology does not imply EPA approval of the conclusions of any specific life-cycle inventory.

Library of Congress Cataloging-in-Publication Data

Catalog information is available from the Library of Congress.

Lewis Publishers is an imprint of CRC Press

International Standard Book Number 1-56670-015-9
Printed in the United States of America 1 2 3 4 5 6 7 8 9 0
Printed on acid-free paper

FOREWORD

Today's rapidly developing and changing technologies and industrial products and practices frequently carry with them the increased generation of materials that, if improperly dealt with, can threaten both public health and the environment. The U.S. Environmental Protection Agency (EPA) is charged by Congress with protecting the Nation's land, air, and water resources. Under a mandate of national environmental laws, the Agency strives to formulate and implement actions leading to a compatible balance between human activities and the ability of natural systems to support and nurture life. These laws direct the EPA to perform research to define our environmental problems, measure the impacts, and search for solutions.

The Risk Reduction Engineering Laboratory is responsible for planning, implementing, and managing research, development, and demonstration programs to provide an authoritative, defensible engineering basis in support of the policies, programs, and regulations of the EPA with respect to drinking water, wastewater, pesticides, toxic substances, solid and hazardous wastes, and Superfund-related activities. This publication is one of the products of that research and provides a vital communication link between the researcher and the user community.

This document is written in a manner to be useful to a broad audience. This audience includes organizations currently conducting studies, those intending to conduct such studies, and those interpreting studies done by other organizations. By providing a template for generalizing the inventory development process and describing a set of rules to assist in making necessary assumptions regarding, for example, assessment boundaries, data quality and coverage, and equivalency of use in a consistent fashion, the guide should reduce the tendency for studies to be published with apparently contradictory conclusions. The added methodological structure should also aid organizations in reading, evaluating, and applying the results of inventories by articulating desired quality milestones.

E. Timothy Oppelt, Director

Risk Reduction Engineering Laboratory

ABSTRACT

This document describes the three components of a life-cycle assessment (inventory analysis, impact analysis, and improvement analysis) as well as scoping activities, presents a brief overview of the development of the life-cycle assessment process, and develops guidelines and principles for implementation of a product life-cycle assessment. The major stages in a life cycle are raw materials acquisition, manufacturing, consumer use/reuse/maintenance, and recycle/waste management. The basic steps of performing a life-cycle inventory (defining the goals and system boundaries, including scoping; gathering and developing data; presenting and reviewing data; and interpreting and communicating results) are presented along with the general issues to be addressed. The system boundaries, assumptions, and conventions to be addressed in each stage of the inventory are presented. Life-cycle impact analysis and life-cycle improvement analysis will be topics of forthcoming guidance documents.

This report was submitted by Battelle in fulfillment of Contract No. 68-C0-0003 under the sponsorship of the U.S. Environmental Protection Agency. Technical effort leading to this report covers a period from August 1990 to May 1991. The draft report was completed in September 1991. Following a comment period, this final report was prepared between March and November 1992.

CONTENTS

Page

Page

Page

LIST OF FIGURES

LIST OF TABLES

ACKNOWLEDGMENTS

This document was prepared for the U.S. Environmental Protection Agency (EPA) Office of Research and Development (ORD), Office of Air Quality Planning and Standards (OAQPS), Office of Solid Waste (OSW), and Office of Pollution Prevention and Toxics (OPPT). Mary Ann Curran of ORD, Risk Reduction Engineering Laboratory (RREL), Cincinnati, Ohio, served as project officer. Additional EPA guidance, reviews, and comments were provided by David C. Fege (OPPT), Eun-Sook Goidel (OPPT), Michael Flynn (OSW), Paul Kaldjian (OSW), Lynda Wynn (OSW), Eugene Lee (OSW), Timothy Mohin (OAQPS), Timothy Ream (OAQPS), and Anne Robertson (RREL). The technical work was conducted under Battelle/EPA Contract No. 68-C0-0003 by Battelle with Bruce Vigon as work assignment leader. Karl Nehring and Steve Pomeroy provided technical contributions. Technical review was also provided by Sid Everett (SRI International). Vincent Brown and Lynn Copley-Graves edited the document and Diane Holbrook coordinated publication.

Peer reviewers included Bob Berkebile, American Institute of Architects; Joel Charm, Allied-Signal, Inc.; Frank Consoli, Scott Paper; Michelle Crew, New Jersey Department of Environmental Protection; Gary Davis, University of Tennessee; Norman Dean, Green Seal, Inc.; Richard Denison, Environmental Defense Fund; Michael Harrass, U.S. Food and Drug Administration; Greg Keoleian, University of Michigan; John Kusz, Industrial Design Society of America; Gail Mayville, Ben & Jerry's; Beth Quay, Coca-Cola U.S.A.; Derek Augood, Scientific Certification Systems, Inc.; T. Michael Rothgeb and Charles Pittinger, Procter & Gamble; Jacinthe Séguin, Environment Canada; Karen Shapiro, Tellus Institute; William W. Walton, U.S. Consumer Product Safety Commission; Matt Weinberg, Office of Technology and Assessment; and Jeanne Wirka, Environmental Action Foundation.

Views contained in this document may not necessarily reflect those of individual reviewers.

SUMMARY AND INDEX OF GUIDING STATEMENTS AND KEY PRINCIPLES

This summary combines the key principles and recommendations from within the body of the report into one list for quick reference. Page numbers indicate the location of each item in the text.

Page

CHAPTER THREE

Page

Chapter One

INTRODUCTION

The concept of life-cycle assessment is to evaluate the environmental effects associated with any given activity from the initial gathering of raw material from the earth until the point at which all residuals are returned to the earth. This concept, often referred to as "cradle to grave" assessment, is not new. While the practice of conducting life-cycle studies has existed for more than 20 years, there has been no comprehensive attempt to describe the procedure in a manner that would facilitate understanding of the overall process, the underlying data, and the inherent assumptions. The literature contains few published assessments and even fewer peer-reviewed publications describing the technical basis for life-cycle assessments. The Society for Environmental Toxicology and Chemistry (SETAC) life-cycle assessment technical framework workshop report published in January 1991 summarizes the current status of the field and outlines the technical basis for life-cycle studies. The purpose of this U.S. Environmental Protection Agency inventory guidelines and principles document is to provide guidance on the specific details involved in the conduct of life-cycle studies.

Some of the most promising applications of life-cycle assessment are for internal use by corporations and regulatory agencies. By developing and using information regarding environmental effects that are both "upstream" and "downstream" of the particular activity under scrutiny, a new paradigm is created for basing decisions in both corporate management and regulatory policy-making.

Recently, there has been a sharp increase in the number of groups conducting life-cycle assessments. Often, the results of these studies have been used to support public claims about various products or processes. Predictably, the results of these studies are often in conflict, and somewhat dependent on the group sponsoring the study. With the increasing use of this information to gain a competitive advantage in the marketplace, there is a clear need for neutral, scientifically oriented, consensus-based guidelines on the conduct of life-cycle assessment.

The EPA has initiated a project to develop such guidelines. The project involves a multi-office EPA group devoted to addressing methodological issues concerning life-cycle assessments. This core group consists of representatives from the Office of Research and Development, Office of Solid Waste, Office of Air Quality Planning and Standards, and Office of Pollution Prevention and Toxics. This inventory guidelines document is the first in a series on conducting life-cycle assessment studies. Additional documents will follow as the knowledge and understanding of life-cycle assessment evolves. Near-term efforts include the preparation of documents that provide guidance on the impact analysis component, on data availability, and on data quality issues for life-cycle assessments. Improvement analysis and guidance for streamlining life-cycle studies are potential future products of this core group.

The EPA's life-cycle assessment project includes using a consensus-building approach and working in close coordination with SETAC. As a scientific and professional society, SETAC has provided infrastructure, credibility, resources, and technical expertise to the development of life-cycle concepts both in the United States and internationally. Through the organization of a series of workshops, SETAC has overseen the development of an emerging technical framework for the conduct of life-cycle assessment.

Based on discussions at the 1990 SETAC workshop, life-cycle assessment consists of three components: inventory analysis, impact analysis, and improvement analysis. This document is intended to be a practical guide to conducting and interpreting life-cycle inventory analysis, which consists of an accounting of the resource usage and environmental releases associated with a product, process, or activity throughout each stage of its life cycle. Recently, the SETAC model has been expanded to include an initial step of goal definition and subsequent scoping analysis. These newer aspects of the SETAC model serve to tailor the scope and boundaries of life-cycle studies to be appropriate with the stated goals of the study. To the extent practicable, this document incorporates the concepts of goal definition and scoping as they apply to life-cycle inventory analysis.

Recent SETAC activity also has begun to define a conceptual framework for life-cycle impact analysis. Preliminary findings of this effort suggest that certain categories of impacts may require expanded or modified inventory data collection. To the extent that these requirements can be anticipated, this document incorporates the additional scope.

This document is not a "cookbook." Given the range of applications, it is not feasible to provide "recipes" for every situation that could be encountered. Instead, this guide attempts to provide a rationale for ensuring internal consistency of procedures for both data acquisition and calculation used in life-cycle inventory analyses. This document relies heavily on practices that have been used by some life-cycle practitioners and that have evolved over many years. Certain decision rules in this guide are presented as specific recommendations because they have proven to be practical over their years of use. In other cases, where judgment is essential regarding an assumption in the study, the guide presents the relevant alternatives with some of the associated advantages and disadvantages. There is full recognition within this guide that the practice of life-cycle assessment continues to evolve. This guidance should be viewed as a starting point, capturing a "snapshot" of the state of the science of life-cycle inventory assessment. As the overall life-cycle assessment framework continues to evolve, it is very likely that changes to the inventory methods presented herein will be necessary.

Currently there is no single correct way to conduct a life-cycle assessment. One clear message of this document is that when a life-cycle practitioner makes assumptions or defines the boundary conditions of a life-cycle study, these decisions must be transparent to the users of that study. In other words, it is imperative for the credibility of the study that the goals, scope, and all assumptions inherent in any life-cycle study are clear to the audience for that study. It is recommended that all groups having a stake in the continued development and application of life-cycle assessment adopt the recommendations contained in this guide and fully disclose all of the assumptions used in the conduct of their life-cycle studies.

The guidance manual consists of five chapters. Chapter Two provide a methodology overview, including the status of current research and the basics of the life-cycle assessment methodology. Readers familiar with the concept may wish to skip this chapter. Subsequent chapters assume a considerable familiarity with the terms and concepts used for life-cycle studies. Chapter Three describes a technical framework for conducting a life-cycle inventory. Chapter Four discusses general issues in performing a life-cycle inventory. Chapter Five contains descriptions and analyses of issues pertaining to the individual stages and steps of a life-cycle inventory: raw materials acquisition; manufacturing (including materials manufacture, product fabrication, and filling/packaging/distribution); consumer use/reuse/maintenance; and recycle/waste management.

Chapter Two

OVERVIEW

LIFE-CYCLE ASSESSMENT CONCEPT

Over the past 20 years, environmental issues have gained greater public recognition. The general public has become more aware that the consumption of manufactured products and marketed services, as well as the daily activities of our society, adversely affect supplies of natural resources and the quality of the environment. These effects occur at all stages of the life cycle of a product, beginning with raw material acquisition and continuing through materials manufacture and product fabrication. They also occur during product consumption and a variety of waste management options such as landfilling, incineration, recycling, and composting. As public concern has increased, both government and industry have intensified the development and application of methods to identify and reduce the adverse environmental effects of these activities.

> **Major Concepts**
>
> - Life-cycle assessment is a tool to evaluate the environmental consequences of a product or activity holistically, across its entire life.
> - There is a trend in many countries toward more environmentally benign products and processes.
> - A complete life-cycle assessment consists of three complementary components: Inventory, Impact, and Improvement Analyses.
> - Life-cycle inventories can be used both internally to an organization and externally, with external applications requiring a higher standard of accountability.
> - Life-cycle inventory analyses can be used in process analysis, material selection, product evaluation, product comparison, and policy-making.

Life-cycle inventory is a "snapshot" of inputs to and outputs from a system. It can be used as a technical tool to identify and evaluate opportunities to reduce the environmental effects associated with a specific product, production process, package, material, or activity. This tool can also be used to evaluate the effects of resource management options designed to create sustainable systems. Life-cycle inventories may be used both internally by organizations to support decisions in implementing product, process, or activity improvements and externally to inform consumer or public policy decisions. External uses are expected to meet a higher standard of accountability in methodology application. Life-cycle assessment adopts a holistic approach by analyzing the entire life cycle of a product, process, package, material, or activity. Life-cycle stages encompass extraction and processing of raw materials; manufacturing, transportation, and distribution; use/reuse/maintenance; recycling and composting; and final disposition. It is not the intent of a life-cycle assessment to analyze economic factors. A life-cycle assessment can be used to create scenarios upon which a cost analysis could be performed.

The three separate but interrelated components of a life-cycle assessment include (1) the identification and quantification of energy and resource use and environmental releases to air, water, and land (inventory analysis); (2) the technical qualitative and quantitative characterization and assessment of the consequences on the environment (impact analysis); and (3) the evaluation and implementation of opportunities to reduce environmental burdens (improvement analysis). Some life-cycle assessment practitioners have defined a fourth component, the scoping and goal definition or initiation step, which serves to tailor the analysis to its intended use.

Life-cycle assessment is not necessarily a linear or stepwise process. Rather, information from any of the three components can complement information from the other two. Environmental benefits can be realized from each component in the process. For example, the inventory analysis alone may be used to identify opportunities for reducing emissions, energy consumption, and material use. The impact analysis addresses ecological and human health consequences and resource depletion, as well as other effects, such as habitat alteration, that cannot be analyzed in the inventory. Data definition and collection to support impact analysis may occur as part of inventory preparation. Improvement analysis helps ensure that any potential reduction strategies are optimized and that improvement programs do not produce additional, unanticipated adverse impacts to human health and the environment. This guidance document is concerned primarily with inventory analyses.

A BRIEF HISTORY OF LIFE-CYCLE INVENTORY ANALYSIS

Life-cycle inventory analysis had its beginnings in the 1960s. Concerns over the limitations of raw materials and energy resources sparked interest in finding ways to cumulatively account for energy use and to project future resource supplies and use. In one of the first publications of its kind, Harold Smith reported his calculation of cumulative energy requirements for the production of chemical intermediates and products at the World Energy Conference in 1963.

Later in the 1960s, global modeling studies published in *The Limits to Growth* (Meadows et al., 1972) and *A Blueprint for*

A Life-Cycle Assessment Has Three Components

These components overlap and build on each other in the development of a complete life-cycle assessment.

- Inventory Analysis
- Impact Analysis
- Improvement Analysis

Scoping is an activity that initiates an assessment, defining its purpose, boundaries, and procecures.

Survival (Club of Rome) resulted in predictions of the effects of the world's changing population on the demand for finite raw materials and energy resources. The predictions of rapid depletion of fossil fuels and climatological changes resulting from excess waste heat stimulated more detailed calculations of energy use and output in industrial processes. During this period, about a dozen studies were performed to estimate costs and environmental implications of alternative sources of energy.

In 1969 researchers initiated a study for The Coca-Cola Company that laid the foundation for the current methods of life-cycle inventory analysis in the United States. In a comparison of different beverage containers to determine which container had the lowest releases to the environment and least affected the supply of natural resources, this study quantified the raw materials and fuels used and the environmental loadings from the manufacturing processes for each container. Other companies in both the United States and Europe performed similar comparative life-cycle inventory analyses in the early 1970s. At this time, many of the data were derived from publicly available sources such as government documents or technical papers, as specific industrial data were not available.

The process of quantifying the resource use and environmental releases of products became known as a Resource and Environmental Profile Analysis (REPA), as practiced in the United States. In Europe it was called an Ecobalance. With the formation of public interest groups encouraging industry to ensure the accuracy of information in the public domain, and with the oil shortages in the early 1970s, approximately 15 REPAs were performed between 1970 and 1975. Through this period, a protocol or standard research methodology for conducting these studies was developed. This multistep methodology involves a number of assumptions. During these years, the assumptions and techniques used underwent considerable review by EPA and major industry representatives, with the result that reasonable methodologies evolved.

From 1975 through the early 1980s, as interest in these comprehensive studies waned because of the fading influence of the oil crisis, environmental concern shifted to issues of hazardous waste management. However, throughout this time, life-cycle inventory analyses continued to be conducted and the methodology improved through a slow stream of about two studies per year, most of which focused on energy requirements. During this time, European interest grew with the establishment of an Environment Directorate (DG X1) by the European Commission. European life-cycle assessment practitioners developed approaches parallel to those being used in the USA. Besides working to standardize pollution regulations throughout Europe, DG X1 issued the Liquid Food Container Directive in 1985, which charged member companies with monitoring the energy and raw materials consumption and solid waste generation of liquid food containers.

When solid waste became a worldwide issue in 1988, the life-cycle inventory analysis technique again emerged as a tool for analyzing environmental problems. As interest in all areas affecting resources and the environment grows, the methodology for life-cycle inventory analysis is again being improved. A broad base of consultants and research institutes in North America and Europe have been further refining and expanding the methodology. With recent emphasis on recycling and composting resources found in the solid waste stream, approaches for incorporating these waste management options into the life-cycle inventory analysis have been developed. Interest in moving beyond the inventory to analyzing the impacts of environmental resource requirements and emissions brings life-cycle assessment methods to another point of evolution.

During the past 2 years, the Society of Environmental Toxicology and Chemistry (SETAC) has served as a focal point for technical developments in the life-cycle assessment arena. Workshops on the overall technical framework, impact analysis, and data quality were held to allow consensus building on methodology and acceptable professional practice. Public forums and a newsletter have provided additional opportunity for input from the user community.

Over the past 20 years, most life-cycle inventories have examined different forms of product packaging such as beverage containers, food containers, fast-food packaging, and shipping containers. Many of these inventories have supported efforts to reduce the amount of packaging in the waste stream or to reduce the environmental emissions of producing the packaging.

Some studies have looked at actual consumer products, such as diapers and detergents, while others have compared alternative industrial processes for the manufacture of the same product.

OVERVIEW OF LIFE-CYCLE ASSESSMENT METHODOLOGY

Three Components

Inventory Analysis

The inventory analysis component is a technical, data-based process of quantifying energy and raw material requirements, atmospheric emissions, waterborne emissions, solid wastes, and other releases for the entire life cycle of a product, package, process, material, or activity. Qualitative aspects are best captured in the impact analysis, although it could be useful during the inventory to identify these issues. In the broadest sense, inventory analysis begins with raw material extraction and continues through final product consumption and disposal. Some inventories may have more restricted boundaries because of their intended use (e.g., internal industrial product formulation improvements where raw materials are identified). Inventory analysis is the only component of life-cycle analysis that is well developed. Its methodology has been evolving over a 20-year period. Refinement and enhancement continue to occur following the SETAC workshop in 1990. Chapters Three through Five present the current framework, assumptions, and steps in inventory analysis.

Impact Analysis

The impact analysis component is a technical, quantitative, and/or qualitative process to characterize and assess the effects of the resource requirements and environmental loadings (atmospheric and waterborne emissions and solid wastes) identified in the inventory stage. Methods for impact analysis are in the early stage of development following a SETAC workshop in early

1992. The analysis should address both ecological and human health impacts, resource depletion, and possibly social welfare. Other effects, such as habitat modification and heat and noise pollution that are not easily amenable to the quantification demanded in the inventory, are also part of the impact analysis component.

The key concept in the impact analysis component is that of stressors. The stressor concept links the inventory and impact analysis by associated resource consumption and releases documented in the inventory with potential impacts. Thus, a stressor is a set of conditions that may lead to an impact. For example, a typical inventory will quantify the amount of SO_2 released per product unit, which then may produce acid rain and which in turn might affect the acidification in a lake. The resultant acidification might change the species composition to eventually create a loss of biodiversity.

An important distinction exists between life-cycle impact analysis and other types of impact analysis. Life-cycle impact analysis does not necessarily attempt to quantify any specific actual impacts associated with a product or process. Instead, it seeks to establish a linkage between the product or process life cycle and potential impacts. The principal methodological issue is managing the increased complexity as the stressor-impact sequence is extended. Methods for analysis of some types of impacts exist, but research is needed for others.

Improvement Analysis

The improvement analysis component of the life-cycle assessment is a systematic evaluation of the needs and opportunities to reduce the environmental burden associated with energy and raw material use and waste emissions throughout the life cycle of a product, process, or activity. This analysis may include both quantitative and qualitative measures of improvements. This component has not been widely discussed in a public forum.

Scoping or Initiation

Scoping is one of the first activities in any life-cycle assessment and is considered by some practitioners as a fourth component. During scoping, the product, process, or activity is defined for the context in which the assessment is being made. The scoping process links the goal of the analysis with the extent, or scope, of the study, i.e., what will or will not be included. For some applications, an impact analysis will be desired or essential. In these cases, the preparation of the inventory is not a standalone activity. The scoping process will need to reflect the intent to define and collect the additional inventory data for the impact analysis.

For internal life-cycle inventories, scoping may be done informally by project staff. Scoping for external studies may require the establishment of a multi-organization group and a formal procedure for reviewing the study boundaries and methodology.

Although scoping is part of life-cycle analysis initiation, there may be valid reasons for reevaluating the scope periodically during a study. As the life-cycle inventory model is defined or as data are collected, scope modifications may be necessary.

Identifying and Setting Boundaries for Life-Cycle Stages

The quality of a life-cycle inventory depends on an accurate description of the system to be analyzed. The necessary data collection and interpretation is contingent on proper understanding of where each stage of a life cycle begins and ends.

General Scope of Each Stage

Raw Materials Acquisition

This stage of the life cycle of a product includes the removal of raw materials and energy sources from the earth, such as the harvesting of trees or the extraction of crude oil. Transport of the raw materials from the point of acquisition to the point of raw materials processing is also considered part of this stage.

Manufacturing

The manufacturing stage produces the product or package from the raw materials and delivers it to consumers. Three substages or steps are involved in this transformation: materials manufacture, product fabrication, and filling/packaging/distribution.

Materials Manufacture. This step involves converting a raw material into a form that can be used to fabricate a finished product. For example, several manufacturing activities are required to produce a polyethylene resin from crude oil: The crude oil must be refined; ethylene must be produced in an olefins plant and then polymerized to produce polyethylene; transportation between manufacturing activities and to the point of product fabrication is considered part of materials manufacture.

Product Fabrication. This step involves processing the manufactured material to create a product ready to be filled or packaged, for example, blow molding a bottle, forming an aluminum can, or producing a cloth diaper.

Filling/Packaging/Distribution. This step includes all manufacturing processes and transportation required to fill, package, and distribute a finished product. Energy and environmental wastes caused by transporting the product to retail outlets or to the consumer are accounted for in this step of a product's life cycle.

Use/Reuse/Maintenance

This is the stage consumers are most familiar with, the actual use, reuse, and maintenance of the product. Energy requirements and environmental wastes associated with product storage and consumption are included in this stage.

Recycle/Waste Management

Energy requirements and environmental wastes associated with product disposition are included in this stage, as well as post-consumer waste management options such as recycling, composting, and incineration.

Stages of a Life Cycle

- Raw Materials Acquisition
- Manufacturing
  - — Materials Manufacture
  - — Product Fabrication
  - — Filling/Packaging/Distribution
- Use/Reuse/Maintenance
- Recycle/Waste Management

Issues That Apply to All Stages

The following general issues apply across all four life-cycle stages:

Energy and Transportation

Process and transportation energy requirements are determined for each stage of a product's life cycle. Some products are made from raw materials, such as crude oil, which are also used as sources of fuel. Use of these raw materials as inputs to products represents a decision to forego their fuel value. The energy value of such raw materials that are incorporated into products typically is included as part of the energy requirements in an inventory analysis. Energy required to acquire and process the fuels burned for process and transportation use is also included.

Environmental Waste Aspects

Three categories of environmental wastes are generated from each stage of a product's life cycle: atmospheric emissions, waterborne wastes, and solid wastes. These environmental wastes are generated by both the actual manufacturing processes and the use of fuels in transport vehicles or process operations.

Waste Management Practices

Depending on the nature of the product, a variety of waste management alternatives may be considered: landfilling, incineration, recycling, and composting.

Allocation of Waste or Energy Among Primary and Coproducts

Some processes in a product's life cycle may produce more than one product. In this event, all energy and resources entering a particular process and all wastes resulting from it are allocated among the product and coproducts. Allocation is most commonly based on the mass ratios of the products, but there are exceptions to this.

Summing the Results of Each Stage

To calculate the total results for the entire life cycle of a particular product, the energy and certain emission values for each stage of the product's life cycle can be summed. For example, energy requirements for each stage are converted from fuel units to million Btus or megajoules and summed to find the total energy requirements. Solid wastes may be summed in pounds or converted to volume and summed. The current, preferred practice is to present the individual environmental releases into each of the environmental media on a pollutant-by-pollutant basis. Where such specificity in an external study would reveal confidential business information, exceptions should be made on a case-by-case basis. Claims for confidentiality should be made only when it is reasonable to expect that release of the information would damage the supplier's competitive position. Even then, the data inputs to an external use are legitimately expected to be independently verified. A peer review process leading to agreed-upon reporting is one possible mechanism for dealing with this issue. Other approaches for independent verification are possible.

APPLICATIONS OF AN INVENTORY ANALYSIS

An inventory conforming to the scope defined in this document will provide a quantitative catalog of energy and other resource requirements, atmospheric emissions, waterborne emissions, and solid wastes for a specific product, process, package, material, or activity. Once an inventory has been performed and is deemed as accurate as possible within the defined scope and boundaries of the system, the results can be used directly to identify areas of greater or lesser environmental burden, to support a subsequent life-cycle impact analysis, and as part of a preliminary improvements analysis. Life-cycle impact assessment can be applied to quantify the human and ecological health consequences associated with specific pollutants identified by the inventory.

The following are possible applications for life-cycle inventories. These are organized according to whether the application is supportable with the inventory alone or whether some level of additional impact analysis is appropriate. The critical issue that users should keep in mind is that if an application context results only in an inventory, the resulting information must not be over-interpreted. Inventories can be applied internally to an organization or externally to convey information outside of the sponsoring organization. External uses are broadly defined in this document to include any study where results will be presented or used beyond the boundaries of the sponsoring organization. Most applications will require some level of impact analysis in addition to the inventory.

To Support Broad Environmental Assessments

The results of an inventory are valuable in understanding the relative environmental burdens resulting from evolutionary changes in given processes, products, or packaging over time; in understanding the relative environmental burdens between alternative processes or materials used to make, distribute, or use the same product; and in comparing the environmental aspects of alternative products that serve the same use.

To Establish Baseline Information

A key application of a life-cycle inventory is to establish a baseline of information on an entire system given current or predicted practices in the manufacture, use, and disposition of the product or category of products. In some cases it may suffice to establish a baseline for certain processes associated with a product or package. This baseline would consist of the energy and resource requirements and the environmental loadings from the product or process systems analyzed. This baseline information is valuable for initiating improvement analysis by applying specific changes to the baseline system.

To Rank the Relative Contribution of Individual Steps or Processes

The inventory provides detailed data regarding the individual contributions of each step in the system studied to the total system. The data can provide direction to efforts for change by showing which steps require the most energy or other resources, or which steps contribute the most pollutants. This application is especially relevant for internal industry studies to support decisions on pollution prevention, resource conservation, and waste minimization opportunities.

These first three applications are supportable with the understanding that the inventory data convey no information as to the possible environmental consequences of the resource use or releases. Any interpretation beyond the "less is best" approach is subjective.

To Identify Data Gaps

The performance of life-cycle inventory analyses for a particular system reveals areas in which data for particular processes or regarding current practices are lacking or are of uncertain or questionable quality. When the inventory is to be followed by an impact analysis, this use can also identify areas where data augmentation for the impact analysis is appropriate.

To Support Policy

For the public policymaker, life-cycle inventories and impact analyses can help broaden the range of environmental issues considered in developing regulations or setting policies.

To Support Product Certification

Product certifications have tended to focus on relatively few criteria. Life-cycle inventories, only when augmented by appropriate impact analyses, can provide information on the individual, simultaneous effects of many product attributes.

To Provide Education for Use in Decision-Making

Life-cycle inventories and impact analyses can be used to educate industry, government, and consumers on the tradeoffs of alternative processes, products, materials, and/or packages. The data can give industry direction in decisions regarding production materials and processes and create a better informed public regarding environmental issues and consumer choices.

These last three applications of life-cycle inventories are the most prone to overinterpretation. This is partly due to their more probable use external to the performing organization and partly due to their implicit orientation towards assessing the environmental consequences of a product or process.

FORMAT OF THIS REPORT

The remainder of the report provides more specific guidance and application examples for those who perform or interpret life-cycle inventories. Chapter Three presents the procedural framework for performing a life-cycle inventory, defines the scope and structure, and describes the construction of the model, the collection and availability of sources of data, and the presentation of results. Chapter Four discusses issues common to all stages of life-cycle inventory analysis. Chapter Five discusses in greater detail issues pertinent to specific life-cycle stages. This guide is not intended to be a point by point prescription for the inventory preparation process. Given the dynamic nature of the science and the developing methods for impact analysis, potential users are provided with some degree of methodological flexibility while still maintaining scientific integrity and transparency.

Chapter Three

PROCEDURAL FRAMEWORK FOR LIFE-CYCLE INVENTORY

INTRODUCTION

This chapter describes the procedural framework for performing a life-cycle inventory. Although there is broad scientific agreement on the major elements of a life-cycle inventory, procedural decisions occur at many steps in the process. This guidance document presents these decisions in the form of a decision tree. The advantages and limitations of each option are discussed.

Major Concepts

- Clear definitions of the purposes and boundaries of a life-cycle inventory analysis help ensure valid interpretation of the results.
- In life-cycle inventory analysis, the term "system" refers to a collection of operations that together perform some well-defined function.
- A broad life-cycle inventory accounts for every significant step in a product system.
- System flow diagrams and calculations are used to determine the resource requirements and environmental emissions for a product.
- Interpretation of results depends on boundary conditions, the quality of data, and the assumptions used.
- A peer review process for external application inventories should be implemented early in the study.

When possible, each option is presented as typical or desired practice based on previous technical forums and scientific soundness.

The inventory process begins with a conceptual goal definition phase to define both the purpose for performing the inventory and the scope of the analysis. An inventory procedure is then employed, and data on the product or system are gathered. Next, the data are incorporated into a computer model to determine the results for the entire system. Additional adjustments in system boundaries and collection of data may be necessary as a result of analyzing preliminary results. Finally, results are presented and interpreted. Chapter Four discusses more general issues related to performing an inventory.

The remaining sections of this chapter describe the steps of performing a life-cycle inventory. They are as follows:

- Define the Purpose and Scope of the Inventory
- Define the System Boundaries
- Devise an Inventory Checklist
- Institute a Peer Review Process
- Gather Data
- Develop Stand-Alone Data

- Construct a Computational Model
- Present the Results
- Interpret and Communicate the Results.

DEFINE THE PURPOSE AND SCOPE OF THE INVENTORY

The decision to perform a life-cycle inventory usually is based on one or several of the following objectives:

- To establish a baseline of information on a system's overall resource use, energy consumption, and environmental loadings
- To identify stages within the life-cycle of a product or process where a reduction in resource use and emissions might be achieved
- To compare the system inputs and outputs associated with alternative products, processes, or activities
- To help guide the development of new products, processes, or activities toward a net reduction of resource requirements and emissions
- To help identify areas to be addressed during life-cycle impact analysis.

These objectives can be further categorized into applications by alternative user groups as listed below.

Private Sector Uses

Evaluation for Internal Decision Making

- Compare alternative materials, products, processes, or activities within the organization.
- Compare resource use and release inventory information with comparable information on other manufacturers' products.
- Train personnel responsible for reducing the environmental burdens associated with products, processes, and activities, including product designers and engineers.
- Provide the baseline information needed to carry out other components of the life-cycle assessment.

Evaluation for Public Disclosure of Information

- Provide information to policymakers, professional organizations, and the general public on resource use and releases, including appropriate disclosure and documentation of findings.
- Help substantiate product-related statements of quantifiable reductions in energy, raw materials, and environmental releases, provided that information is not selectively reported.

Public Sector Uses

Evaluation and Policymaking

- Supply information for evaluating existing and prospective policies that affect resource use and releases.
- Develop policies and regulations on materials and resource use and environmental releases when the inventory is supplemented by an impact analysis.
- Identify gaps in information and knowledge, and help establish research priorities and monitoring requirements on the state and federal levels.

- Evaluate product statements of quantifiable reductions in energy, raw materials, and environmental releases.

Public Education

- Develop materials to help the public understand resource use and release characteristics associated with products, processes, and activities.
- Design curricula for training those involved in product, process, and activity design.

Key decisions in defining the scope and boundaries of the inventory rest on the defined goal or purpose of the inventory. Conceptually, it is useful to distinguish between the study boundaries and the system boundaries. Study boundaries include both the issues to be dealt with and the physical system boundaries to be analyzed.

Study Specificity

At the outset of every study, the level of specificity must be decided. In some cases, this level will be obvious from the application or intended use of the information. In other instances, there may be several options to choose among, ranging from a completely generic study to one that is product-specific in every detail. Most studies fall somewhere in between.

A life-cycle inventory can be envisioned as a set of linked activities that describe the creation, use, and ultimate disposition of the product or material of interest. At each life-cycle stage, the analyst should begin by answering a series of questions: Is the product or system in this life-cycle stage specific to one company or manufacturing operation? Or does the product or system represent common products or systems generally found in the marketplace and produced or used by a number of companies?

Such questions help determine whether data collected for the inventory should be specific to one company or manufacturing facility, or whether the data should be more general to represent common industrial practices.

The appropriate response to these questions often rests on whether the life-cycle inventory is being performed for internal organizational use or for a more public purpose. Accessibility to product- or facility-specific data may also be a factor. A company may be more interested in examining its own formulation and assembly operations, whereas an industry group or government agency may be more interested in characterizing industry-wide practice. Life-cycle inventories can have a mix of product-specific and industry-average information. For example, a cereal manufacturer performing an analysis of using recycled paperboard for its cereal boxes might apply the following logic. For operations conducted by the manufacturer, such as box printing, setup, and filling, data specific to the product would be obtained because average data for printing and filling across the cereal industry or for industry in general would not be as useful.

Stepping back one stage to package manufacturing, the cereal manufacturer is again faced with the specificity decision. The inventory could be product-specific, or generic data for the package manufacturing stage could be used. The product-specific approach has these advantages: the aggregated inventory data will reflect the operations of the specific papermills supplying the recycled board, and the energy and resources associated with this stage can be compared with those of similar specificity for the filling, packaging, and distribution stage. A limitation of this option is the additional cost and time associated with collecting product-specific data from the mills and the level of cooperation that needs to be established with the upstream vendors. Long-term confidentiality agreements with vendors may also represent unacceptable burdens compared with the value added by the more specific data.

The alternative decision path, using industrial average data for making recycled paperboard, has a parallel mix of advantages and limitations. Use of average, or generic, data may be advantageous for a manufacturer considering use of recycled board for which no current vendors have been identified. If the quality of these average data can be determined and is acceptable, their use may be preferable. The limitation is that data from this stage may be less comparable to that of more product-specific stages. This limitation is especially important in studies that mix product-specific and more general analyses in the same life-cycle stage. For example, comparing virgin and recycled paperboard using product-specific data for one material and generic data for the other could be problematic.

Another limitation is that the generic data may mask technologies that are more environmentally burdensome. Even with some measure of data variability, a decision to use a particular material made on the basis of generic data may misrepresent true loadings of the actual suppliers. Opportunities to identify specific facilities operating in a more environmentally sound manner are lost. Generic data do not necessarily represent industry-wide practices. The extent of representation depends on the quality and coverage of the available data and is impossible to state as a general rule.

It is recommended that the level of specificity be very clearly defined and communicated so that readers are more able to understand differences in the final results. Before initiating data collection and periodically throughout the study, the analyst should revisit the specificity decision to determine if the approach selected for each stage remains valid in view of the intended use.

DEFINE THE SYSTEM BOUNDARIES

Once the goal or purpose for preparing a life-cycle inventory has been determined and the intended use is known, the system should be specifically defined. A "system" is a collection of operations that together perform some clearly defined function. A broad-based system begins with raw materials acquisition and continues through industrial or consumer use and final disposition. Great care should be taken in defining the systems to be analyzed and in explaining the boundaries for the definitions in any report of inventory results. Clear definitions help ensure valid interpretations of the results.

In defining the system, the first step is to set the system boundaries. A complete life-cycle inventory will set the boundaries of the total system broadly to quantify resource and energy use and environmental releases throughout the entire life cycle of a product or process, as shown in Figure 1. The life-cycle stages used in this guidance document differ in two respects from those presented in the SETAC Technical Framework (Fava et al., 1991). These differences arise because this guidance document is written from the perspective of assembling information to perform an inventory study. For this reason, transportation activities used to move materials from one stage to the next have been disaggregated rather than being presented as a separate stage. Each transportation step is associated with a specific upstream life-cycle stage. Although it is common to present transportation-related energy and emissions separately in reporting results, the transportation system type and the distances covered are defined within each stage.

This model combines materials manufacture, product fabrication, and filling/packaging/distribution in the manufacturing stage. Separating these three aspects of manufacturing into separate substages, or steps, reflects the fact that different organizations typically are involved in these activities. The separate treatment also reflects the different nature of the operations and the decisions discussed earlier regarding product, material, or activity specificity. Figure 2 shows the three steps of manufacturing.

Recycling and waste management are combined into one stage because, especially for post-consumer material, recycling and

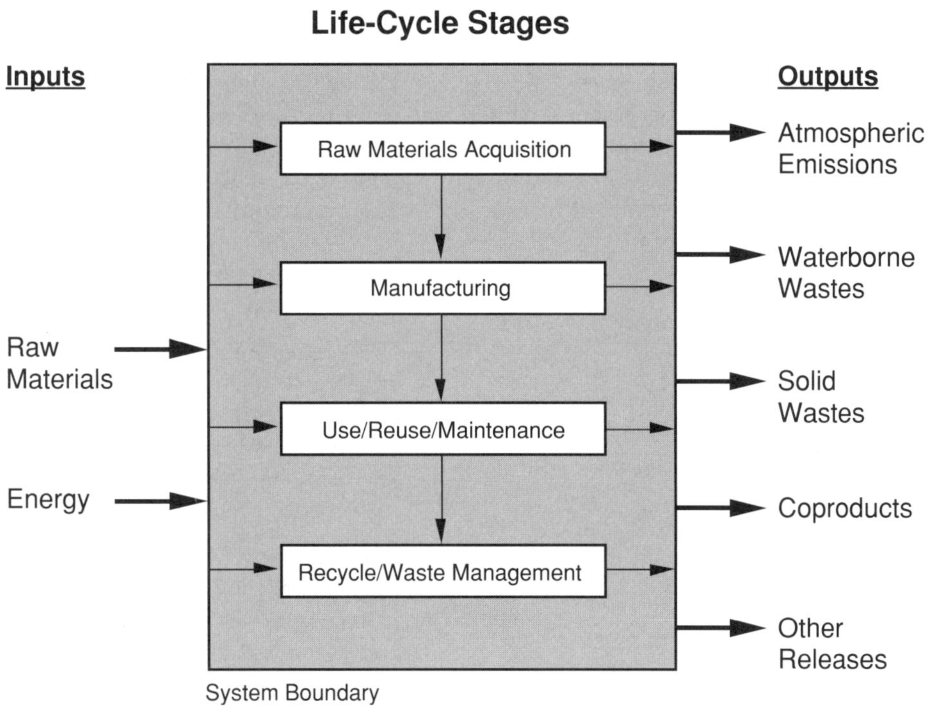

Figure 1
Defining system boundaries

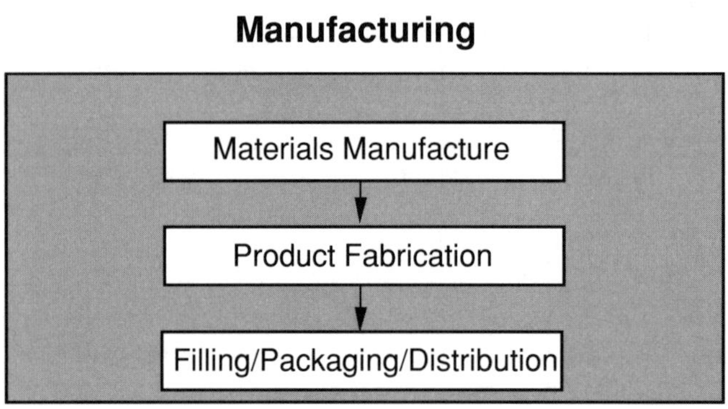

Figure 2
Steps in the manufacturing stage

waste management simply represent a splitting of the material flow between the two streams. A life cycle, therefore, comprises four major stages:

- **Raw Materials Acquisition Stage.** All the activities required to gather or obtain a raw material or energy source from the earth. This stage includes transportation of the raw material to the point of material manufacture, but does not include material processing activities.
- **Manufacturing Stage.** Encompasses three steps:
  - **Materials Manufacture.** The activities required to process a raw material into a form that can be used to fabricate a particular product or package. Normally, the production of many intermediate chemicals or materials is included in this category. Transport of intermediate materials is also included.
  - **Product Fabrication.** The process step that uses raw or manufactured materials to fabricate a product ready to be filled or packaged. This step often involves a consumer product that will be distributed for retail sales, but the product could also be distributed for use by other industries.
  - **Filling/Packaging/Distribution.** Processes that prepare the final products for shipment and that transport the products to retail outlets. Although these activities may commonly require a change in the location or physical configuration of a product, they do not involve a transformation of materials.
- **Use/Reuse/Maintenance Stage.** Begins after the distribution of products or materials for intended use and includes any activity in which the product or package may be reconditioned, maintained, or serviced to extend its useful life.
- **Recycle/Waste Management Stage.** Begins after the product, package, or material has served its intended purpose and either will enter a new system through recycling or will enter the environment through the waste management system.

Each step in the life cycle of a product, package, or material can be categorized within one and only one of these life-cycle stages. Each step or process can be viewed as a subsystem of the total product system. Viewing the steps as subsystems facilitates data gathering for the inventory of the system as a whole. The boundaries of subsystems are defined by life-cycle stage categories in Chapter Five. The rest of this chapter deals with defining the boundaries of the whole product system. Many decisions must be made in defining the specific boundaries of each system.

Product systems are easier to define if the sequence of operations associated with a product or material is broken down into primary and secondary categories. The primary, or zero-order, sequence of activities directly contributes to making, using, or disposing of the product or material. The secondary category includes auxiliary materials or processes that contribute to making or doing something that in turn is in the primary activity sequence. Several tiers of auxiliary materials or processes may extend further and further from the main sequence. In setting system boundaries, the analyst must decide where the analysis will be limited and be very clear about the reasons for the decision. The following questions are useful in setting and describing specific system boundaries:

- Does the analysis need to cover the entire life cycle of the product? A theoretically complete life-cycle system would start with all raw materials and energy sources in the earth and end with all materials back in the earth or at least somewhere in the environment but not part of the system. Any system boundary different from this represents a decision by the analyst to limit it in some way. Understanding the possible consequences of such decisions is important for evaluating tradeoffs between the ability of the resulting inventory to thoroughly address environmental attributes of the product and constraints on cost, time, or other factors that may argue in favor of a more limited boundary. Too limited a boundary may exclude consequential activities or elements.

Depending on the goal of the study, it may be possible to exclude certain stages or activities and still address the issues for which the life-cycle inventory is being performed. For example, it may be possible to exclude the acquisition of raw materials in a life-cycle inventory without affecting the results. Suppose a company wishes to perform an internal life-cycle inventory to evaluate alternative drying systems for formulating a snack food product. If the technologies are indifferent to feedstock, it is possible to assume the raw materials acquisition stage will be identical for all options. If the decision will be based on selecting a drying system with lower energy use or environmental burdens, it may be acceptable to analyze such a limited system. However, with this system boundary, the degree of absolute difference in the overall system energy or environmental inventory cannot be determined. The difference in the product manufacturing stage, although significant for the manufacturer, may represent a minor component of the total system. Therefore, statements about the total system should not be made.

- What will be the basis of use for the product or material? Is the study intended to compare different product systems? If the products or processes are

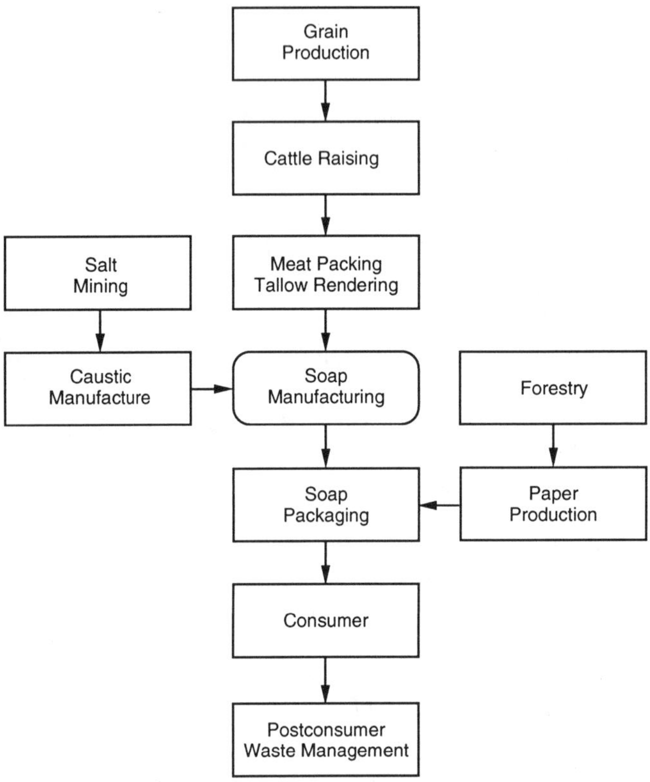

Figure 3
Example system flow diagram for bar soap

used at different rates, packaged in varying quantities, or come in different sizes, how can one accurately compare them? Can equivalent use ratios be developed? Should market shares be considered to estimate the proportionate burden from each product in a given category? Is the study intended to compare service systems? Are the service functions clearly defined so that the inputs and outputs are properly proportioned?

- What ancillary materials or chemicals are used to make or package the products or run the processes? Might these ancillary materials or chemicals contribute more than a minor fraction of the energy or emissions of the system to be analyzed? How do they compare by weight with other materials and chemicals in the product system?
- In a comparative analysis, are any extra products required to allow one product to deliver equivalent or similar performance to another? Are any extra materials or services required for one service to be functionally equivalent to another or to a comparable product?

Figure 3 shows an example of setting system boundaries for a product baseline analysis of a bar soap system. Tallow is the major material for soap production, and its primary raw material source is the grain fed to cattle. Production of paper for packaging the soap is also included. The fate of both the soap and its packaging end the life cycle of this system. Minor inputs could include, for example, the energy required to fabricate the tires on the combine used to plant and harvest the grain.

In a life-cycle inventory to create a baseline for future product development or improvement, the unit upon which the analysis is performed can be almost anything that produces internally consistent data. In the bar soap example, one possible usage unit could be a single bar. However, if the product packaging were being analyzed at the same time, it would be important for consistency to consider packaging in different amounts such as single bars, three packs, and so on.

If the life-cycle inventory were intended to analyze whether bar soap should be manufactured using an animal-derived or vegetable-derived raw material source, the system boundaries and units of analysis would be more complicated. First, the system flow diagram would have to be expanded to include the growing, harvesting, and processing steps for the alternative feedstock. Then the performance of the finished product would have to be considered. Do the options result in a bar that gets used up at different rates when one material or the other is chosen? If this were the case, a strict comparison of equal-weight bars would not be appropriate.

Suppose an analyst wants to compare bar soap made from tallow with a liquid hand soap made from synthetic ingredients. Because the two products have different raw material sources (cattle and petroleum), the analysis should begin with the raw materials acquisition steps. Because the two products are packaged differently and may have different chemical formulas, the materials manufacture and packaging steps would need to be included. Consumer use and waste management options also should be examined because the different formulas could result in varying usage patterns. Thus for this comparative analysis, the analyst would have to inventory the entire life cycle of the two products.

Again, the analyst must determine the basis of comparison between the systems. Because one soap is a solid and the other is a liquid, each with different densities and cleansing abilities per unit amount, it would not make sense to compare them based on equal weights or volumes. The key factor is how much of each is used in one hand-washing to provide an equivalent level of function or service. An acceptable basis for comparison might be equal numbers of hand-washings. Because these two products may be used at different rates, it would be important to find data that give an equivalent use ratio. For example, a research lab study may show that 5 mm^3 of bar soap and 10 mm^3 of liquid soap are used per hand-washing. If the basis for comparison were chosen at 1,000 hand-washings, 5,000 mm^3 of bar soap would be compared to 10,000 mm^3 of liquid hand soap. Thus, the equivalent use ratio is 1 to 2.

When specific brands of soap are being analyzed to establish individual life-cycle inventories, market share information need not be obtained. However, when two or more items perform the same function and the inventory application is framed to answer the question of identifying which product type exhibits higher or lower loadings, market share information is important in comparing product types. For example, in an analysis comparing a typical bar soap with a typical liquid soap, market share data would be used to allocate the raw materials and emissions among the specific soap brands of each type. These data would be used to proportion the contributions of chemicals, raw materials, usage rates, etc., of the two soap types to develop average data. On the other hand, if specific soap brands are being compared, then, of course, market share is also irrelevant.

Because the two soap product types are packaged in different quantities and materials, the analyst would need to include packaging in the system. Contributions of extra ingredients, such as perfumes, might also be considered. The analyst may or may not find that any extra raw materials are used in one of the two soap types to make it clean as well as the other. Soaps typically must meet a minimum standard performance level.

However, if the liquid hand soap also had a skin moisturizer in its formula, the analyst would need to include a moisturizing lotion product in the boundary of the bar soap system on two conditions. The first condition would apply if the environmental issues associated with this component were germane to the purpose of the life-cycle inventory. The second condition, which is not as clear-cut, is if there is actual value received by the consumer from inclusion of the moisturizer. If market studies indicate that consumers purchase the product in preference to an identical product without a moisturizer, or if they subsequently use a moisturizing lotion after using a nonmoisturizing soap, then equivalent use would entail including the separate moisturizing lotion. Including the moisturizing lotion would move the comparison beyond equivalent hand-washing to equivalent hand-washing and skin moisturizing.

In defining system boundaries, it is important to include every step that could affect the overall interpretation or ability of the analysis to address the issues for which it is being performed. Only in certain well-defined instances can life-cycle elements such as raw materials acquisition or waste management be excluded. In general, only when a step is exactly the same in process, materials, and quantity in all alternatives considered, can that step be excluded from the system. In addition, the framework for the comparison must be recognized as relative because the total system values exclude certain contributions. This rule is especially critical for inventories used in public forums rather than for internal company decision making. For example, a company comparing alternative processes for producing one petrochemical product may not need to consider the use and disposal of the product if the final composition is identical. The company may also find that each process uses exactly the same materials in the same amounts per unit of product output. Therefore, the company may consider the materials it uses as having no impact on the study results. Another example is a filling operation for bottles. A company interested in using alternative materials for its bottles while maintaining the same size and shape may not need to include filling the bottles as part of the inventory system. However, if the original bottles were compared to boxes of a different size and shape, the filling step would need to be included.

When a life-cycle inventory is used to compare two or more products, the basis of comparison should be equivalent use; i.e., each system should be defined so that an equal amount of product or equivalent service is delivered to the consumer. In the hand-washing example, if bar soap were compared to liquid soap, the logical basis for comparison would be an equal number of hand-washings. Another example of equivalent use would be in comparing cloth diapers to disposable diapers. One type of diaper may typically be changed more frequently than the other, and market/use studies show that often cloth diapers are doubled, whereas disposables are not. Thus, throughout a day, more cloth diapers will be used. In this case, a logical basis for comparison between the systems would be the total number of diapers used over a set period of time.

Equivalent use for comparative studies can often be based on volume or weight, particularly when the study compares packaging for delivery of a specific product. A beverage container study might consider 1,000 liters of beverage as an equivalent use basis for comparison, because the product may be delivered to the consumer in a variety of different-size containers having different life-cycle characteristics.

Resource constraints for the life-cycle inventory may be considerations in defining the system boundaries, but in no case should the scientific basis of the study be compromised. The level of detail required to perform a thorough inventory depends on the size of the system and the purpose of the study. In a large system encompassing several industries, certain details may not be significant contributors given the defined intent of the study. These details may be omitted without affecting the accuracy or application of the results. However, if the study has a very specific focus, such as a manufacturer comparing alternative processes or materials for inks used on packaging, it would be important to include chemicals used in very small amounts.

Additional areas to consider in setting boundaries include the manufacture of capital equipment, energy and emissions associated with personnel requirements, and precombustion impacts for fuel usage. These are discussed in Chapter Four.

After the boundaries of each system have been determined, a system flow diagram as shown in Figure 3 can be developed to depict the system. Each system step should be represented individually in the diagram, including the production steps for ancillary inputs or outputs such as chemicals and packaging. If a decision to exclude certain items has been made, it is appropriate, for purposes of maintaining transparency, to explain the system flow diagram. Often in studies intended for external application, the system flow diagram is incorporated into a formal scope, boundary, and data collection (SBDC) document. An SBDC document provides the basis for peer reviewers and others to understand how the analyst is defining the study and how these definitions and assumptions will translate into data collection activities.

DEVISE AN INVENTORY CHECKLIST

The inventory checklist is a tool that covers most decision areas in the performance of an inventory. After the inventory purpose and boundaries have been defined, a checklist can be prepared to guide data collection and validation and to enable construction of the computational model. Figure 4 shows a generic example of an inventory checklist and an accompanying data worksheet. Although this checklist is an effective guidance tool and enhances transparency, it is not the sole quality control process under which the analysis should be performed. Analysts will want to tailor this checklist for a given product or material. Eight general decision areas should be addressed on the checklist or worksheet:

- Purpose of the inventory
- System boundaries
- Geographic scope

LIFE-CYCLE INVENTORY CHECKLIST PART I—SCOPE AND PROCEDURES
INVENTORY OF: ______________________________

Purpose of Inventory: *(check all that apply)*

Private Sector Use

Internal Evaluation and Decision Making
- 1 Comparison of Materials, Products, or Activities
- 1 Resource Use and Release Comparison with Other Manufacturer's Data
- 1 Personnel Training for Product and Process Design
- 1 Baseline Information for Full LCA

External Evaluation and Decision Making
- 1 Provide Information on Resource Use and Releases
- 1 Substantiate Statements of Reductions in Resource Use and Releases

Public Sector Use

Evaluation and Policy-making
- 1 Support Information for Policy and Regulatory Evaluation
- 1 Information Gap Identification
- 1 Help Evaluate Statements of Reductions in Resource Use and Releases

Public Education
- 1 Develop Support Materials for Public Education
- 1 Assist in Curriculum Design

Systems Analyzed

List the product/process systems analyzed in this inventory: ______________________________

____

Key Assumptions: (list and describe)

____

____

____

Define the Boundaries

For each system analyzed, define the boundaries by life-cycle stage, geographic scope, primary processes, and ancillary inputs included in the system boundaries.

Postconsumer Solid Waste Management Options: Mark and describe the options analyzed for each system.
- 1 Landfill ______________________________
- 1 Combustion ______________________________
- 1 Composting ______________________________
- 1 Open-loop Recycling ______________________________
- 1 Closed-loop Recycling ______________________________
- 1 Other ______________________________

Basis for Comparison

1 This is not a comparative study. 1 This is a comparative study.

State basis for comparison between systems: *(Example: 1000 units, 1,000 uses)* ______________________________

____

If products or processes are not normally used on a one-to-one basis, state how equivalent function was established.

____

Computational Model Construction
- 1 System calculations are made using computer spreadsheets that relate each system component to the total system.
- 1 System calculations are made using another technique. Describe: ______________________________

____

Describe how inputs to and outputs from postconsumer solid waste management are handled. ______________________________

____

____

Quality Assurance: (state specific activities and initials of reviewer)

Review performed on:
- 1 Data Gathering Techniques ______________________________
- 1 Coproduct Allocation ______________________________
- 1 Input Data ______________________________
- 1 Model Calculations and Formulas ______________________________
- 1 Results and Reporting ______________________________

Peer Review: (state specific activities and initials of reviewer)

Review performed on:
- 1 Scope and Boundary ______________________________
- 1 Data Gathering Techniques ______________________________
- 1 Coproduct Allocation ______________________________
- 1 Input Data ______________________________
- 1 Model Calculations and Formulas ______________________________
- 1 Results and Reporting ______________________________

Results Presentation
- 1 Methodology is fully described.
- 1 Individual pollutants are reported.
- 1 Emissions are reported as aggregrated totals only. Explain why: ______________________________

  ____
- 1 Report is sufficiently detailed for its defined purpose.
- 1 Report may need more detail for additional use beyond defined purpose.
- 1 Sensitivity analyses are included in the report. List: ______________________________
- 1 Sensitivity analyses have been performed but are not included in the report. List: ______________________________

  ____

Figure 4

A typical checklist of criteria with worksheet for performing a life-cycle inventory

LIFE-CYCLE INVENTORY CHECKLIST PART II—MODULE WORKSHEET

Inventory of: ______________________ Preparer: ______________

Life-Cycle Stage Description: ______________________

Date: ______________ Quality Assurance Approval: ______________

MODULE DESCRIPTION: ______________________

	Data Value[(a)]	Type[(b)]	Data[(c)] Age/Scope	Quality Measures[(d)]
MODULE INPUTS				
Materials				
Process				
Other[(e)]				
Energy				
Process				
Precombustion				
Water Usage				
Process				
Fuel-related				
MODULE OUTPUTS				
Product				
Coproducts[(f)]				
Air Emissions				
Process				
Fuel-related				
Water Effluents				
Process				
Fuel-related				
Solid Waste				
Process				
Fuel-related				
Capital Repl.				
Transportation				
Personnel				

(a) Include units.

(b) Indicate whether data are actual measurements, engineering estimates, or theoretical or published values and whether the numbers are from a specific manufacturer or facility, or whether they represent industry-average values. List a specific source if pertinent, e.g., "obtained from Atlanta facility wastewater permit monitoring data."

(c) Indicate whether emissions are all available, regulated only, or selected. Designate data as to geographic specificity, e.g., North America, and indicate the period covered, e.g., average of monthly for 1991.

(d) List measures of data quality available for the data item, e.g., accuracy, precision, representativeness, consistency-checked, other, or none.

(e) Include nontraditional inputs, e.g., land use, when appropriate and necessary.

(f) If coproduct allocation method was applied, indicate basis in quality measures column, e.g., weight.

Figure 4
Continued

- Types of data used
- Data collection and synthesis procedures
- Data quality measures
- Computational model construction
- Presentation of the results.

A standard checklist can be helpful in several settings. The analyst performing the life-cycle inventory can use the checklist to help ensure that all important stages and categories of information are included. The checklist can also help clarify the issues, boundaries, and conditions to be dealt with in a particular study. Worksheets can be used by the analyst to collect and qualify data from facilities.

The checklist consists of two major components—a summary section describing the procedures and systems included in the study and a set of worksheets listing and qualifying the data collected. The checklist is to be used throughout a study to ensure that all of the boundaries and comparison issues are identified and addressed during the study. The worksheet portion has a dual purpose: as a tool for the analyst to coordinate and assimilate data and for use in requesting data from others. In a life-cycle inventory where there may be many steps in each life-cycle stage, the worksheets help ensure consistency among the various information sources. Modules consisting of subsystem inputs and outputs are the basis for preparing a life-cycle inventory. Subsystem modules represent fundamental operations that are building blocks for aggregating data to the life-cycle stage and overall system level. For example, a module may be constructed for the production of caustics as shown in the system flow diagram for bar soap (Figure 3). This module would be one of several comprising the materials manufacture substage. Worksheets may be prepared for each facility in each subsystem module and may be used by the analyst to aggregate these data to the life-cycle stage level. Additional internal quality control and quality assurance procedures should be in place to ensure that the inventory is complete and sound.

A checklist such as the one in Figure 4 also can be used as a communication tool. By including a completed checklist in the report on the results of an inventory, the analyst can communicate to readers some of the factors that may affect the results. The checklist will help readers gain knowledge and understanding of the system's boundaries, data quality, methodology used, and level of detail. Standard checklists included in reports of inventories also can help readers recognize and understand differences among various reports on the same topic. The standard checklist also can be of use to peer reviewers as it provides useful criteria and information to assure the completeness of a particular life-cycle inventory.

INSTITUTE A PEER REVIEW PROCESS

The desirability of a peer review process has been a major focus of discussion in many life-cycle analysis forums. The discussion stems from concerns in four areas: lack of understanding regarding the methodology used or the scope of the study, desire to verify data and the analyst's compilations of data, questioning key assumptions and the overall results, and communication of results. For these reasons, it is recommended that a peer review process be established and implemented early in any study that will be used in a public forum.

SETAC is working with business, consumer, and environmental groups, and with academia to develop a peer review process for inventory studies (Fava et al., 1992). The following discussion is *not* intended to be a blueprint of a specific approach. Instead, it is meant to point out issues that the practitioner or sponsor should keep in mind when establishing a peer review procedure.

Overall, a peer review process should address the four areas previously identified:

- Scope/boundaries methodology
- Data acquisition/compilation
- Validity of key assumptions and results
- Communication of results.

The peer review panel could participate at several points in the study: (1) reviewing the purpose, system boundaries, assumptions, and data collection approach; (2) reviewing the compiled data and the associated quality measures; and (3) reviewing the draft inventory report, including the intended communication strategy.

A checklist such as the one presented in Figure 4 would be useful in addressing many of the issues surrounding scope/ boundaries methodology, data/compilation of data, and validity of assumptions and results. Criteria may need to be established for communication of results. These criteria could include showing how changes in key assumptions could affect the study results and guidance on how to publish and communicate results without disclosing proprietary data.

It is generally believed that the peer review panel should consist of a diverse group of 3 to 5 individuals representing various sectors, such as federal, state, and local governments; academia; industry; environmental or consumer groups; and LCA practitioners. Not all sectors need be represented on every panel. The credentials or background of individuals should include a reputation for objectivity, experience with the technical framework or conduct of life-cycle studies, and a willingness to work as part of a team. Issues for which guidelines are still under development include panel selection, number of reviews, using the same reviewers for all life-cycle studies or varying the members between studies, and having the review open to the public prior to its release. The issue of how the reviews should be performed raises a number of questions, such as these: Should a standard checklist be required? Should oral as well as written comments from the reviewers be accepted? How much time should be allotted for review? Who pays for the review process?

The peer review process should be flexible to accommodate variations in the application or scope of life-cycle studies. Peer review should improve the conduct of these studies, increase the understanding of the results, and aid in further identifying and subsequently reducing any environmental consequences of products or materials. EPA supports the use of peer reviews as a mechanism to increase the quality and consistency of life-cycle inventories.

GATHER DATA

The system flow diagram (Figure 3) is useful in conjunction with the checklist and worksheets (Figure 4) in directing efforts to gather data for the life-cycle inventory.

Identify Subsystems

For data-gathering purposes it is appropriate to view the system as a series of subsystems. A "subsystem" is defined as an individual step or process that is part of the defined production system. Some steps in the system may need to be grouped into a subsystem due to lack of specific data for the individual steps. For example, several steps may be required in the production of bar soap from tallow. However, these steps may all occur within the same facility, which may not be able to or need to break data down for each individual step. The facility could, however, provide data for all the steps together, so the subsystem boundary would be drawn around the group of soap production steps and not around each individual one.

Each subsystem requires inputs of materials and energy; requires transportation of product produced; and has outputs of products, coproducts, atmospheric emissions, waterborne wastes, solid wastes, and possibly other releases. For each subsystem, the inventory analyst should describe materials and energy sources used and the types of environmental releases. The actual activities that occur should also be described. Data should be gathered for the amounts and kinds of material inputs and the types and quantities of energy inputs. The environmental releases to air, water, and land should be quantified by type of pollutant. Data collected for an inventory should always be associated with a quality measure. Although formal data quality indicators (DQIs) such as accuracy, precision, representativeness, and completeness are strongly preferred, a description of how the data were generated can be useful in judging quality. EPA is specifically addressing the use of quantitative and qualitative DQIs in a separate guidance document on data quality in life-cycle assessment.

Coproducts from the process should be identified and quantified. Coproducts are process outputs that have value, i.e., those not treated as wastes. The value assigned to a coproduct may be a market value (price) or may be imputed. In performing coproduct allocation, some means must be found to objectively assign the resource use, energy consumption, and emissions among the coproducts, because there is no physical or chemical way to separate the activities that produce them. Advantages and disadvantages of specific approaches to coproduct allocation are discussed later in this chapter and in Chapter Four. Generally, allocation should allow technically sound inventories to be prepared for products or materials using any particular output of a process independently and without overlap of the other outputs.

In the meat packing step of the bar soap example shown in Figure 3, several coproducts could be identified: meat, tallow, bone meal, blood meal, and hides. Other examples of coproducts are the trim scraps and off-spec materials from a molded plastic plate fabricator. If the trim scraps and off-spec materials are used or marketed to other manufacturers, they are considered as coproducts. Industrial scrap is the common name given to such materials. If the trim is discarded into the solid waste stream to be landfilled, it should be included in the solid waste from the process. If the trim or off-spec materials are reused within the process, they are considered "home scrap," which is part of an internal recycling loop. These materials are not included in the

> **Sources of Data**
>
> - Electronic non-bibliographic data bases (government and industrial)
>   - averaged industrial data
>   - product specifications
> - Electronic bibliographic data bases
> - Electronic database clearinghouses
> - Relevant documents
>   - government reports
>   - open literature papers and books
>   - other life-cycle inventories
> - Facility-specific industrial data
>   - publicly accessible
>   - nonpublicly accessible
> - Laboratory test data
> - Study-specific data

inventory, because they do not cross the boundaries of the subsystem.

All transportation from one process location to another is included in the subsystem. Transportation is quantified in terms of distance and weight shipped, and identified by the mode of transport used.

Sources of Data

A number of sources should be used in collecting data. Whenever possible, it is best to get well-characterized industry data for production processes. Manufacturing processes often become more efficient or change over time, so it is important to seek current data. Inventory data can be facility-specific or more general and still remain current.

Several categories of data are often used in inventories. Starting with the most disaggregated, these are:

- Individual process- and facility-specific: data from a particular operation within a given facility that are not combined in any way
- Composite: data from the same operation or activity combined across locations
- Aggregated: data combining more than one process operation
- Industry-average: data derived from a representative sample of locations and believed to statistically describe the typical operation across technologies
- Generic: data whose representativeness may be unknown but which are qualitatively descriptive of a process or technology.

Complete and thorough inventories often require use of data considered proprietary by either the manufacturer of the product, upstream suppliers or vendors, or the LCA practitioner performing the study. Confidentiality issues are not relevant for life-cycle inventories conducted by companies using their own facility data for internal purposes. However, the use of proprietary data is a critical issue in inventories conducted for external use and whenever facility-specific data are obtained from external suppliers for internal studies. As a consequence, current studies often contain insufficient source and documentation data to permit technically sound external review. Lack of technically sound data adversely affects the credibility of both the life-cycle inventories and the method for performing them. An individual company's trade secrets and competitive technologies must be protected. When collecting data (and later when reporting the results), the protection of confidential business information should be weighed against the need for a full and detailed analysis or disclosure of information. Some form of selective confidentiality agreements for entities performing life-cycle inventories, as well as formalization of peer review procedures, is often necessary for inventories that will be used in a public forum. Thus, industry data may need to undergo intermediate confidential review prior to becoming an aggregated data source for a document that is to be publicly released.

The purpose, scope, and boundary of the inventory help the analyst determine the level or type of information that is required. For example, even when the analyst can obtain actual industry data, in what form and to what degree should the analyst show the data (e.g., the range of values observed, industry averages, plant-specific data, best available control techniques)? These questions or decisions can usually be answered if the purpose or scope has been well defined. Typically, most publicly available life-cycle documents present industry averages, while many internal industrial studies use plant-specific data. Recommended practice for external life-cycle inventory studies includes the provision of a measure of data variability in addition to averages. Frequently the measure of variability will be a statistical parameter, such as a standard deviation. Other options, which may be useful for small data sets or where confidentiality may be breached by a reported standard deviation, are discussed in more detail in Chapter Four.

Examples of private industry data sources include independent or internal reports, periodic measurements, accounting or engineering reports or data sets, specific measurements, and machine specifications. One particular issue of interest in considering industrial sources, whether or not a formal public data set is established, is the influence of industry and related technical associations to enhance the accuracy, representativeness, and currentness of the collected data. Such associations may be willing, without providing specific data, to confirm that certain data (about which their members are knowledgeable) are realistic.

Government documents and data bases provide data on broad categories of processes and are publicly available. Most government documents are published on a periodic basis, e.g., annually, biennially, or every 4 years. However, the data published within them tend to be at least several years old. Furthermore, the data found in these documents may be less specific and less accurate than industry data for specific facilities or groups of facilities. However, depending on the purpose of the study and the specific data objectives, these limitations may not be critical. All studies should note the age of the data used. Some useful government documents include:

- U.S. Department of Commerce, Census of Manufacturers
- U.S. Bureau of Mines, Census of Mineral Industries
- U.S. Department of Energy, Monthly Energy Review
- U.S. Environmental Protection Agency, Toxic Release Inventory (TRI) Database.

Government data bases include both non-bibliographic types where the data items themselves are contained in the data base and bibliographic types that consist of references where data may be found.

Technical books, reports, conference papers, and articles published in technical journals can also provide information and data on processes in the system. Most of these are publicly available. Data presented in these sources are often older, and they can be either too specific or not specific enough. Many of these documents give theoretical data rather than real data for processes. Such data may not be representative of actual processes or may deal with new technologies not commercially tested. In using the technical data sources in the following list, the analyst should consider the date, specificity, and relevancy of the data:

- *Encyclopedia of Chemical Technology,* Kirk-Othmer
- Periodical technical journals such as *Journal of the Water Environment Federation*
- Proceedings from technical conferences
- Textbooks on various applied sciences.

Surveys designed to capture information on a representative sample of end users can provide current information on the parameters of product or service use. Surveys typically center around a question:

- How long or how many times is a product or service used before it is discarded (e.g., the number of years a television set has been in use and is expected to be in use)?
- What other materials and what quantities of these materials are used in conjunction with product use or maintenance (e.g., moisturizing lotion use after hand-washing)?
- How frequent is the need for product repair or maintenance (e.g., how often is an appliance repaired over its lifetime, and who does the repair)?
- What other uses does the product have beyond its original purpose?
- What does the end user do with the product when he or she is through with it?

Frequently, the end user will not be able to supply specific information on inputs and outputs. However, the end user can provide data on user practices from which inputs and outputs can be derived. Generally, the end user can be the source of related information from which the energy, materials, and pollutant release inventory can be derived. (An exception would be an institutional or commercial end user who may have some information on energy consumption or water effluents.) Market research firms can often provide qualitative and quantitative usage and customer preference data without the analyst having to perform independent market surveys.

Recycling provides an example of some of the strengths and limitations encountered in gathering data. For some products, economic-driven recycling has been practiced for many years, and an infrastructure and markets for these materials already exist. Data are typically available for these products, including recycling rates, the consumers of the reclaimed materials, and the resource requirements and environmental releases from the recycling activities (collection and reprocessing). Data for materials currently at low recycling rates with newly forming recycling infrastructures are more difficult to obtain. In either case, often the best source for data on resource requirements and environmental releases is the processors themselves. For data on recycling rates and recycled material, consumers and processors may be helpful, but trade associations as well as the consumers of the recycled materials can also provide data. For materials that are recycled at low rates, data will be more difficult to find.

Two other areas for data gathering relate to the system as a whole and to comparisons between and among systems. It is necessary to obtain data on the weights of each component in the product evaluated, either by obtaining product specifications from the manufacturer or by weighing each component. These data are then used to combine the individual components in the overall system analysis. Equivalent use ratios for the products compared can be developed by surveying retailers and consumers, or by reviewing consumer or trade association periodicals.

DEVELOP STAND-ALONE DATA

"Stand-alone data" is a term used to describe the set of information developed to standardize or normalize the individual subsystem module inputs and outputs for the specific product, process, or activity being analyzed. Stand-alone data must be developed for each subsystem to fit the subsystems into a single system. There are two goals to achieve in this step: (1) presenting data for each subsystem consistently by reporting the same product output from each subsystem; (2) developing the data in terms of the life cycle of only the product being examined in the inventory. First, a standard unit of output must be determined for each subsystem. All data could be reported in terms of the production of a certain number of pounds, kilograms, or tons of subsystem product. For example, the harvesting of trees, production of paper, and packaging of soap are all steps in packaging soap as seen in Figure 3. Data for these steps could be developed on the basis of 1,000 tons of output: 1,000 tons of harvested trees, 1,000 tons of paper, and 1,000 tons of packaged soap. Although historically English units have been used for subsystem accounting in the USA, international practice is leaning more toward the use of metric units of measurement. The units used for the individual steps or subsystems do not necessarily have to match those of the final product.

Once the decision for the reporting basis has been made, the data obtained for the subsystem need to be adjusted to that product's output level. For example, suppose an analyst performing a study on bar soap has received data on the caustic manufacture subsystem. This process yields three coproducts: caustic, chlorine, and hydrogen. Assume the data obtained for the process were given in terms of the joint production of 500 tons of caustic, 250 tons of chlorine, and 5 tons of hydrogen. Only the caustic is used in the bar soap production system. Because the analyst has chosen 1,000 tons as the reporting basis for all subsystem data, the caustic manufacturing data need to be presented in terms of 1,000 tons of output of caustic (and thus also 500 tons of chlorine and 10 tons of hydrogen). To do this, all of the process data supplied would be multiplied by two. For purposes of this example, only the chlorine coproduct is considered further.

Now that the data are at a consistent level of reporting, the analyst needs to determine the energy and material requirements and the environmental releases to be attributed to the production of each coproduct using a technique called coproduct allocation. One commonly used allocation method is based on relative weight. Figure 5 illustrates this technique. In this example, the top portion of the figure can be used to illustrate the actual process flow diagram for hypothetical production of caustic (labeled Product A), with chlorine (Product B) as a coproduct of caustic. Because the bar soap system exclusive of packaging uses only caustic, and not chlorine, the energy and material inputs and environmental releases of the process must be allocated separately to the caustic and chlorine. The lower portion of the diagram illustrates this allocation. Product A, caustic, represents two-thirds of the total production output of the process, so two-thirds of the energy and resource inputs and two-thirds of the environmental releases are attributed to caustic. Likewise, Product B, chlorine, represents one-third of the total production output of the process, so one-third of the energy and resource inputs and one-third of the environmental releases are attributed to the chlorine. Performing data allocation in this way allows the analyst to isolate those inputs and outputs relevant to the product being studied. Alternative allocation methods are discussed in Chapter Four.

Once the inputs and outputs of each subsystem have been allocated, numerical relationships of the subsystems within the entire system flow diagram can be established. This is done starting at the finished product of the system and working backward, using the relationships of the material inputs and product outputs of each subsystem to compute the input requirements from each of the preceding subsystems. For example, suppose the bar soap system were to be analyzed on the basis of 1,000 tons of packaged bar soap. If the bar soap packaging process requires 900 tons of bar soap to produce 1,000 tons of packaged bar soap, only 900 tons of bar soap would need to be manufactured for the total system. Suppose 2,000 tons of tallow are required to produce 1,000 tons of bar soap. Only 900 tons of bar soap are required for the total bar soap system, so only 1,800 tons of tallow are needed for the total system.

CONSTRUCT A COMPUTATIONAL MODEL

The next step in a life-cycle inventory is model construction. This step consists of incorporating the normalized data and material flows into a computational framework using a computer spreadsheet or other accounting technique. The systems accounting data that result from the computations of the model give the total results for the energy and resource use and environmental releases from the overall system.

The overall system flow diagram, derived in the previous step, is important in constructing the computational model because it numerically defines the relationships of the individual subsystems to each other in the production of the final product. These numerical relationships become the source of "proportionality factors," which are quantitative relationships that reflect the relative contributions of the subsystems to the total system. For example, data for the production of a particular ingredient X of bar soap are developed for the production of 1,000 tons of X. To produce 1,000 tons of bar soap, 250 tons of X are needed, accounting for losses and inefficiencies. Thus, to find the contributions of X to the total system, the data for 1,000 tons of X are multiplied by 0.250.

Actual process flow diagram for the production of Products 'A' and 'B'

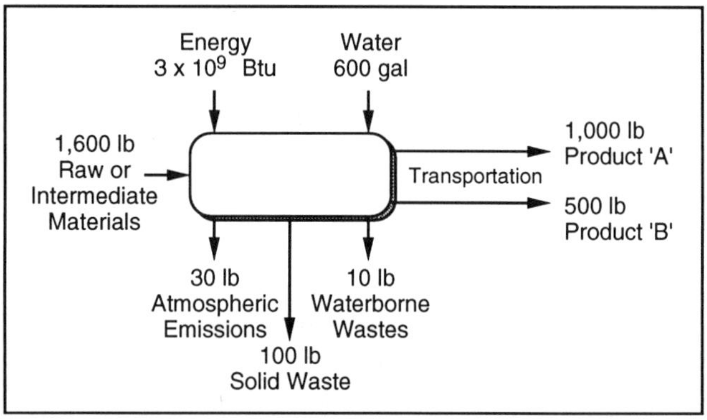

Flow diagrams showing the normalized resources and environmental releases for each coproduct

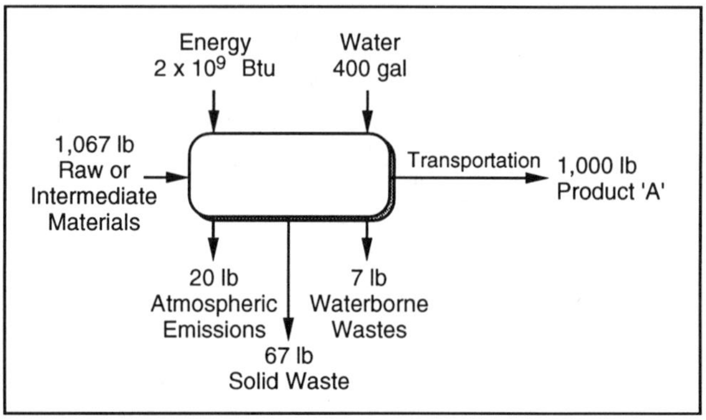

Coproduct Allocation for Product 'A'

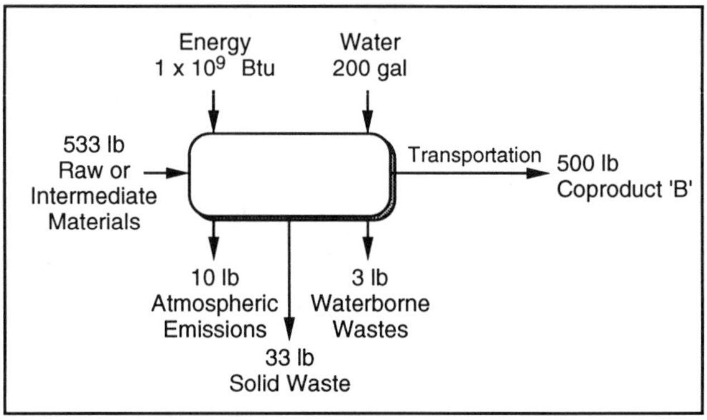

Coproduct Allocation for Product 'B'

Figure 5
Example coproduct allocation based on relative weight

The spreadsheet can be used to make other computations beyond weighting the contributions of various subsystems. It can be used to translate energy fuel value to a standard energy unit, such as million Btu or gigajoule (GJ). Precombustion or resource acquisition energy can be computed by applying a standard factor to a unit quantity of fuel to account for energy used to obtain and transport the fuel. Energy sources, as well as types of wastes, can be categorized. Credits or charges for incineration can be derived. Fuel-related wastes should also be calculated based on the fuels used throughout the system. The model should also incorporate waste management options, such as recycling, composting, and landfilling. The method for handling these aspects is discussed in Chapter Five.

It is important that each subsystem be incorporated in the model with its related components and that each be linked together in such a way that inadvertent omissions and double-counting do not occur. The computer spreadsheet can be organized in several different ways to accomplish this purpose. These can include allocating certain fields or areas in the spreadsheet to certain types of calculations or using one type of spreadsheet software to actually link separate spreadsheets in hierarchical fashion. It is imperative, however, once a system of organization is used, that it be employed consistently. Haphazard organization of data sets and calculations generally leads to faulty inventory results.

Many decisions must be made in every life-cycle inventory analysis. Every inventory consists of a mix of factual data and assumptions. Assumptions allow the analyst to evaluate a system condition when factual data either cannot be obtained within the context of the study or do not exist. Each piece of information (e.g., the weight of paperboard used to package the soap, type of vehicle and distance for shipping the tallow, losses incurred when rendering tallow, or emissions resulting from the animals at the feedlot), falls into one or the other category and each plays a role in developing the overall system analysis. Because assumptions can substantially affect study results, a series of "what if" calculations or sensitivity analyses are often performed on the results to examine the effect of making changes in the system. A sensitivity analysis will temporarily modify one or more parameters and affect the calculation of the results. Observing the change in the results will help determine how important the assumptions are with respect to the results. The computational model is also used to perform these sensitivity analysis calculations.

Sometimes it is helpful to think ahead about how the results will be presented. This can direct some decisions on how the model output is specified. The analyst must remember the defined purpose for performing the analysis and tailor the data output to those expressed needs. For example, the analyst might ask: Is the purpose of the life-cycle inventory to evaluate the overall system results? Or is it expected that detailed subsystem information will be analyzed in relation to the total? Will the study be used in a public forum? If so, how? How much detail is required? Answers to questions such as these will help determine the complexity and the degree of generalization to build into the model, as well as the appropriate presentation of results.

PRESENT THE RESULTS

When writing a report to present the final results of the life-cycle inventory, it is important to thoroughly describe the methodology used in the analysis. The report should explicitly define the systems analyzed and the boundaries that were set. All assumptions made in performing the inventory should be clearly explained. The basis for comparison among systems should be given, and any equivalent usage ratios that were used should be explained. Use of the checklist and worksheet, as shown in Figure 4, supports a clear process for communicating this information.

Life-cycle inventory studies generate a great deal of information, often of a disparate nature. The analyst needs to select a presentation format and content that are consistent with the purpose of the study and that do not arbitrarily simplify the information solely for the sake of presenting it. In thinking about presentation of the results, it is useful to identify the various perspectives embodied in life-cycle inventory information. These dimensions include but may not be limited to the following:

- Overall product system
- Relative contribution of stages to the overall system
- Relative contribution of product components to the overall system
- Data categories within and across stages, e.g., resource use, energy consumption, and environmental releases
- Data parameter groups within a category, e.g., air emissions, waterborne wastes, and solid waste types
- Data parameters within a group, e.g., sulfur oxides, carbon dioxide, chlorine, etc.
- Geographic regionalization if relevant to the study, e.g., national versus global
- Temporal changes.

The life-cycle analyst must select among these dimensions and develop a presentation format that increases comprehension of the findings without oversimplifying them. Two main types of format for presenting results are tabular and graphical.

Sometimes it is useful to report total energy results while also breaking out the contributions to the total from process energy and energy of material resource. Solid wastes can be separated into postconsumer solid waste and industrial solid waste. Individual atmospheric and water pollutants should be reported separately. Atmospheric emissions, waterborne wastes, and industrial solid wastes can also be categorized by process emissions/wastes and fuel-related emissions/wastes. Such itemized presentations can assist in identifying and subsequently controlling certain energy consumption and environmental releases.

The results from the inventory can be presented most comprehensively in tabular form. The choice of how the tables should be created varies, based on the purpose and scope of the study. If the inventory has been performed to help decide which type of package to use for a particular product, showing the overall system results will be the most useful way to present the data. On the other hand, when an analysis is performed to determine how a package can be changed to reduce its releases to the environment, it is important to present not only the overall results, but also the contributions made by each component of the packaging system. For example, in analyzing a liquid delivery system that uses plastic bottles, it may be necessary to show how the bottle, the cap, the label, the corrugated shipping box, and the stretch wrap around the boxes all contribute to the total results. The user can thus concentrate improvement efforts on the components that make a substantial contribution when evaluating proposed changes.

Graphical presentation of information helps to augment tabular data and can aid in interpretation. Both bar charts (either individual bars or stacked bars) and pie charts are valuable in helping the reader visualize and assimilate the information from the perspective of "gaining ownership or participation in life-cycle assessment" (Werner, 1991). However, the analyst should not aggregate or sum dissimilar data when creating or simplifying a graph.

For internal industrial use by product manufacturers, pie charts showing a breakout by raw materials, process, and use/disposal have been found useful in identifying waste reduction opportunities.

For external studies, the data must be presented in a format that meets one fundamental criterion—clarity. Ensuring clarity requires that the analyst ask and answer questions about what each graph is intended to convey. It may be necessary to present a larger number of graphs and incorporate fewer data in each one. Each reader should understand the desired response after viewing the information.

INTERPRET AND COMMUNICATE THE RESULTS

How the results of the life-cycle inventory will be interpreted depends on the purpose for which the analysis was performed. Before publishing any statements regarding the results of the analysis, it is important to review how the assumptions and boundaries of the system were defined, the quality level of the data used, and the specificity (e.g., were the data specific to one facility or representative of the entire industry?). Careful interpretation is required to avoid making unsupported statements.

An important criterion in understanding or interpreting the results is data accuracy. Many life-cycle inventories present data that are considered representative of the industry or group being profiled. Data for one particular step may be gathered from a number of manufacturers and production facilities. For example, for the paperboard manufacturing step, more than 25 plants may produce a similar product. These plants may use different raw materials, employ different technologies, have varying degrees of plant age/efficiency, and operate under different site-specific conditions (e.g., energy sources and state environmental regulations). These aggregated and composited data will be subject to both random and systematic sources of error. Individual process variations within a given facility using the same input and technology represent a source of true random error. This type of variability can be described in conventional statistical terms using the mean and standard deviation of the measurements. For small data sets, where reporting of a mean, range, and standard deviation may compromise confidentiality, a semi-qualitative description of variability could suffice.

However, differences resulting from systematic variability due to feedstock or technology type are not random errors in a statistical sense. These sources of variability may be thought of as "explainable" by the age or operating conditions of the plants or by other identifiable factors.

The analyst should interpret the importance of these sources of variability for the reader. For example, variability analysis for the paperboard manufacturing step could indicate that plants using a certain type of pulping process were consistently higher in certain water pollutants. The analyst could interpret this fact for the reader and still protect confidentiality.

Although a rigorous statistical analysis of the overall inventory may not be possible, sensitivity analyses of key elements in the system should be performed to estimate the magnitude of variability in the data from the inventory that would have to exist in order to reflect significant differences in the results. Sensitivity analysis is a technique for systematically varying the inputs to a model in order to establish whether the outputs are distinguishable or not (Raiffa, 1968). For life-cycle inventories, a sensitivity analysis would evaluate how large the uncertainty in the input data can be before the results of the inventory can no longer be used for the intended purpose.

Typically, a difference in past inventories has been determined to occur where the outputs vary by more than $\pm$10-25 percent when data were modified to simulate low and high ranges. The differences that must exist to use the results will be defined for each inventory during the scoping process as part of the data quality objectives (DQO) setting. The guidelines for performing a sensitivity analysis in a life-cycle inventory are presented more fully in Chapter Four. This margin of difference should be used independently for each category of results (e.g., energy, atmospheric emissions, waterborne emissions, and solid waste) and for each data parameter (e.g., chloroform, particulates, and hydrocarbons) in analyzing the results.

The boundaries and data for many internal life-cycle assessments may require the interpretation of the results for use within a particular corporation. For example, data used may be specific to a particular company and, therefore, may not represent any typical or average product on the market. However, because the data used in this type of analysis are frequently highly specific, a fairly high degree of accuracy can be assumed in interpreting the results. Product design and facility/process development groups within companies often benefit from this level of interpretation.

The public sometimes receives interpreted information from life-cycle analyses that have been released. In these instances, life-cycle inventory information may be provided to consumers to support statements about certain features or specific reduction claims. The analyst must be careful to provide an interpretive context and not to selectively use information. In both types of studies, general systems are described and the data used in the analysis may not be specific to one producer. Instead it may be more representative of a particular industry. In such cases, a higher margin of difference should be achieved before results are considered to be significantly different between systems.

The results of externally published studies comparing products, practices, or materials should be presented cautiously, and assumptions, boundaries, and data quality should be considered in drawing and presenting conclusions. Studies with different boundary conditions may have different results, yet both may be accurate. These limitations should be communicated to the public along with all the results; it is misleading to selectively report inventory results. Final conclusions about the results of inventory studies may involve making value judgments regarding the relative importance of air and water quality, solid waste issues, resource depletion, and energy use. Based upon each individual's locale, background, and life style, different value judgments will be made.

Chapter Four

GENERAL ISSUES IN PERFORMING A LIFE-CYCLE INVENTORY

INTRODUCTION

This chapter discusses the general issues encountered in every stage of a life-cycle inventory. These issues pertain to the type of information these studies quantify and the decisions or assumptions that must be made in evaluating and using the information. One major tool in life-cycle inventory analysis is the template, which is a pictorial guide that identifies the information that must be obtained at each step in an inventory analysis. Issues discussed in this chapter include:

- Using Templates in Life-Cycle Inventory Analysis
- Data Issues
- Special Case Boundary Issues.

Major Concepts

- Templates, or material and energy balance diagrams, are tools used to support data gathering and development for life-cycle inventory analyses.
- Data for processes producing more than one product are allocated based on the relative weights of product output or another justifiable method.
- Data quality objectives are the required performance specifications for information in a life-cycle inventory. Establishment of these specifications is determined by the defined purpose of the life-cycle inventory.
- Data quality indicators are qualitative or quantitative characteristics of data. These include accuracy, bias, representativeness, and other attributes that measure data goodness and applicability.

USING TEMPLATES IN LIFE-CYCLE INVENTORY ANALYSIS

A template is a guide used by analysts to direct the collection of data. A template depicts the material and energy accounts to visually describe a defined system or subsystem. A generic version of a template, shown in Figure 6, indicates which categories of information are necessary to construct the energy and materials input-output analysis that is at the core of traditional life-cycle inventories.

Inputs *(Requirements)*	**Outputs** *(Environmental releases and products)*
Raw or intermediate materials	Atmospheric emissions
Energy	Waterborne wastes
Water	Solid wastes
	Other releases
Other inputs	Products
	Coproducts

The concept of life-cycle studies has been extended in recent years beyond merely specifying physical quantities to a more comprehensive characterization. To accommodate this extension, the life-cycle analyst may want to augment the traditional template with additional information. The data and information that describe the more extensive inputs or outputs should be based on the study purpose, with care taken to ensure that the life-cycle inventory remains a data-driven accounting procedure. The additional data could include categories of information not traditionally considered in a material balance, such as noise, aesthetics, and odors, and more broadly interpreted emissions such as workplace releases and land use changes. Figure 7 illustrates the steps or processes included in a life-cycle inventory for the bar soap system. Individual steps within this overall flow diagram are referred to as subsystems. The bar soap production step is one example of a subsystem. All subsystem steps, including those for secondary/tertiary packaging, together form the bar soap system. The template can be applied at either the subsystem or system level.

As the bar soap production subsystem illustrates, numerous processes or subprocesses are included such as making fatty acids, vacuum distillation, making toilet soap, cutting, and drying. In some cases, the template will be used for data gathering at the subsystem level without evaluating the processes or subprocesses within the subsystem. The decision to gather data at the subsystem level will depend on the nature

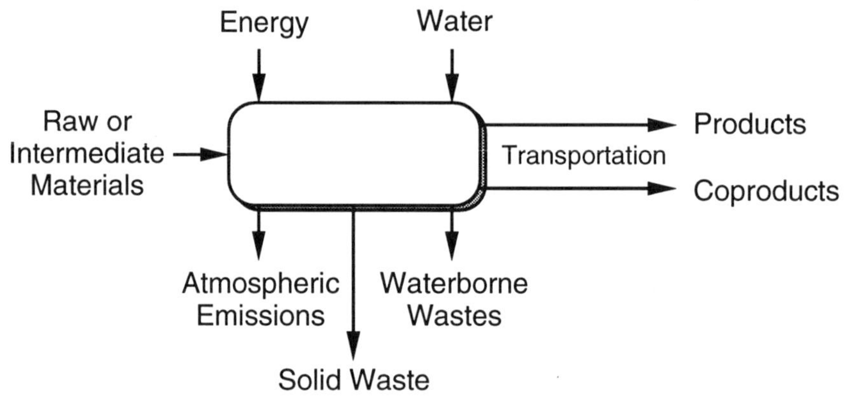

Figure 6
Life-cycle inventory template

Source: Franklin Associates, Ltd.

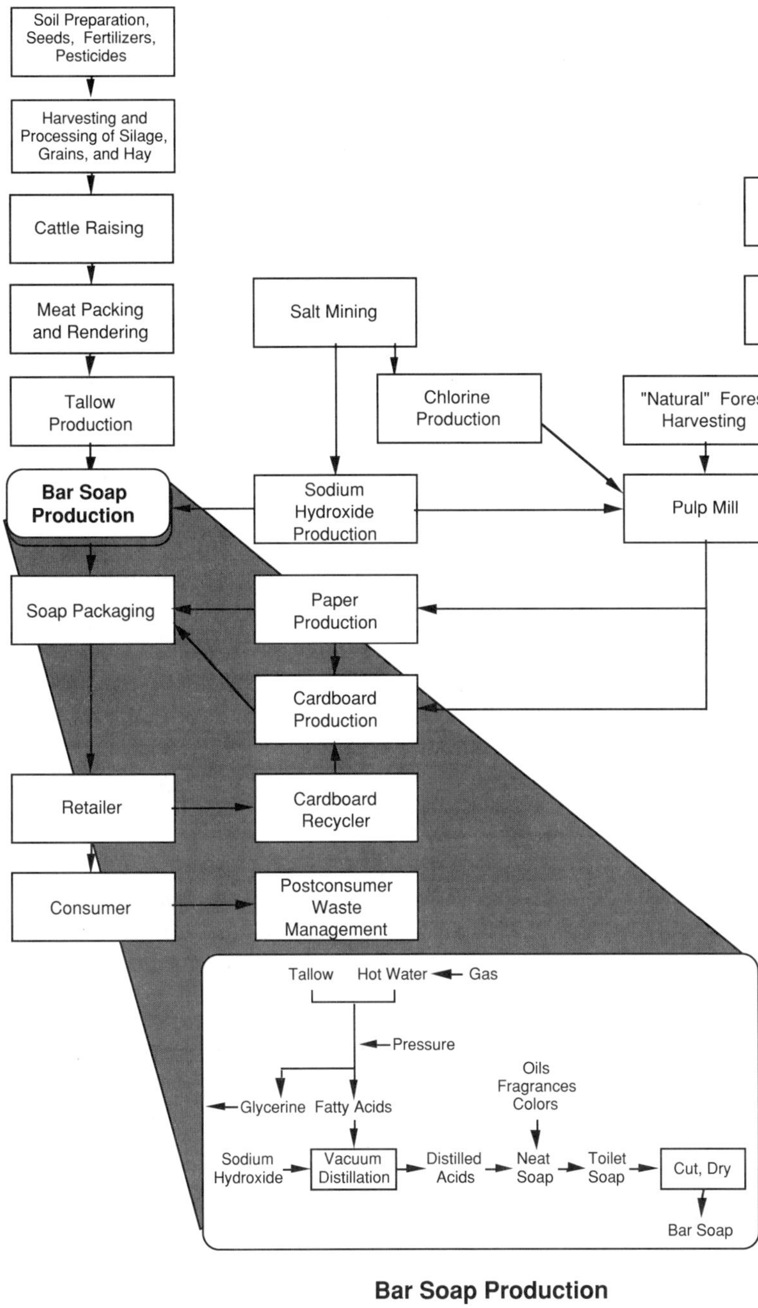

Note: Energy acquisition and electricity generation are not shown on this diagram, although they are inputs to many of these processes.

Figure 7
Detailed system flow diagram for bar soap

of the subsystem and the availability of data. If a specific manufacturing facility has total energy and emissions data for the subsystem, but not specific process data, then it makes sense to apply the template to the entire manufacturing facility (e.g., the bar soap facility in Figure 7) and not to each individual process. In other cases these individual processes must be examined separately, then totaled to provide the subsystem inventory. The following sections describe each of the template components and identify how current life-cycle studies apply criteria to each of these areas.

Inputs in the Product Life-Cycle Inventory Analysis

Raw/Intermediate Materials

The input materials for each subsystem are referred to as raw/intermediate materials. Raw materials are materials that have been extracted from the earth but have not been refined or processed. Intermediate materials are products of one refining or manufacturing step that are input materials into another process. The most complete inventory will begin with all raw materials at the input of the most upstream stage. Any inventory that sets a different boundary should provide the associated reasons and justification for excluding steps. Figure 7 shows that tallow and sodium hydroxide are the two intermediate materials needed for bar soap production, whereas salt is a raw material input for sodium hydroxide production. The raw/intermediate material requirements for a subsystem represent the total material inputs, including all material present in the product and material found in losses from emissions, scrap, and off-spec products as well as non-emission losses (such as moisture due to evaporation). The template is used to guide the documentation of material requirements in pounds or kilograms per unit of product output for each subsystem.

Raw/intermediate materials have been distinguished from other process inputs by some practitioners because raw/intermediate materials are present in the finished product, although they may be transformed chemically. For example, tallow, oils, fragrances, and colors are input materials for bar soap production. Water does not always appear as a raw/intermediate material input. Water may be reclaimed during a drying step even if it is used in the process and, therefore, may not be present in the finished product.

The decision on which raw/intermediate material requirements to include in a life-cycle inventory is complex, but several options are available:

- Incorporate all requirements, no matter how minor, on the assumption that it is not possible, a priori, to decide to exclude anything.
- Within the defined scope of the study, exclude inputs of less than a predetermined and clearly stated threshold.
- Within the defined scope of the study, exclude inputs determined likely to be negligible, relative to the intended use of the information, on the basis of a sensitivity analysis.
- Within the defined scope, consistently exclude certain classes or types of inputs, such as capital equipment replacement.

The advantage of the first option is that no assumptions are made in defining and drawing the system boundary. The analyst does not have to explain or defend what has been included or excluded. The disadvantage is that application of this approach could be an endless exercise. The number of inputs could be very large and could include some systems only distantly related to the product system of interest. Besides the computational complexity, the interpretation of the results with respect to the single desired product, package, or activity could be difficult.

The second option, if implemented with full explanation of what the threshold is and why it was selected, would have the advantages of consistency and lower cost and time investments. Two suboptions can be identified, depending on the nature of the threshold. One suboption is to specify a percentage contribution below which the material will be excluded, for example, 1% of the input to a given subsystem or to the entire system. The 1% rule historically has been useful in limiting the extent of the analysis in inventories where the environmental consequences of quantitatively minor materials are not considered. The disadvantage of the 1% rule is that the possible presence of an environmentally damaging activity associated with these materials could be overlooked. Also, when used with mixed percentages (e.g., percent of system energy, percent of subsystem input), the results may be confusing or inconsistent. The scoping analysis should provide a rationale for choosing to apply such a rule.

The second suboption is to set a threshold based on the number of steps that the raw/intermediate material is removed from the main process sequence. Consider the bar soap example discussed earlier. Caustic manufacture from brine electrolysis is part of the main process sequence and would clearly be included. Sodium carbonate is an input material for the production of caustic and is therefore a secondary input. Applying a "one-step back" decision rule would include the steps associated with sodium carbonate production. Ammonium chloride is an input material for the production of sodium carbonate using the Solvay process. Relative to caustic, ammonium chloride is a tertiary input and would be excluded if a "one-step back" decision rule were applied. As in the first option, the "one-step back" decision rule has the advantages of clarity and consistent application. For some inputs that are analyzable in exact mathematical terms, the "one-step back" rule may be justifiable. If the inputs to a given process bear a fixed relationship to the next-tier process, one step is all that may be necessary to obtain a sufficiently accurate value (Boustead and Hancock, 1979).

Consider the example of a refinery. Most of the refinery's output is sold for production of petroleum-based materials. However, a small portion, say 8%, is used to run the refinery. This portion, termed the parasitic fraction, is mathematically related to the refinery output as:

$$M(1 + f)$$

where:

M is the output product and
f is the parasitic fraction (0.08).

For a life-cycle inventory on a petroleum-based plastic, the primary output of the refinery clearly would be included within the system boundary. Suppose the data quality indicators showed that the data were accurate to ±5%. Because the first-tier use of the material represents an 8% difference, a "one-step back" rule would include the refinery material (fuel) output used to run the refinery. However, to produce the material (fuel) to run the refinery requires a further fraction of the output two steps back from the plastic raw material. This is calculated as:

$$M(1 + f + f^2).$$

Thus, the incremental contribution of the second step back is 0.6%, which is less than the data accuracy. That is, there is no significant difference in the system data after the first step. Disadvantages of this approach include the lack of simple geometric relationships for many inputs and the increased effort to analyze more tiers as data quality increases.

The third option, drawing boundaries based on sensitivity analysis, adds the advantage of being systematic rather than arbitrary in assigning the threshold. The disadvantages of a sensitivity analysis-based approach are that the analyst needs to be very clear in describing how the analysis was used and, unless a large existing database is available to supply preliminary values that can be used in the sensitivity analysis, the required analysis effort may not be limited by a very large amount. A more in-depth discussion of sensitivity analysis is provided later in this chapter.

The final option, excluding certain classes or types of input, also has been found through experience to apply to many systems. For example, in the bar soap inventory, a decision may be made to exclude the equipment used to cut the bars of soap. The justification is that the allocation of inputs and outputs from the manufacture of the machine is miniscule when the millions of bars of soap produced by the machine are considered. The advantage of this option is that many complex subsystems can often be excluded. The disadvantages are the same as those for the first option, namely, that a highly significant activity may be eliminated. Capital equipment is the most commonly excluded input type. The analyst should perform a preliminary analysis to characterize the basic activities in each class or type of input to ensure that a significant contribution is not left out.

Renewable and Nonrenewable Resources

A renewable resource is one that is being replaced in the environment in a time frame relevant to society. Certain species of wood are examples of such a material. Most minerals and metals, along with certain biological products, are of such a stable nature that their supply does not change within a time perspective of several generations, and they are thus deemed nonrenewable (Jorgensen and Pedersen, 1990). Hydrocarbon fuels produced from geological repositories—coal, petroleum, and natural gas—are nonrenewable because they are not being created in a time frame of relevance to humans. If production of hydrocarbon fuels from biomass were to become a significant source of fuels, the definition would be applied to the non-fossil-sourced fraction as well as to that from geological sources.

The intrinsic nature of a resource does not dictate its renewability. A further definition of a sustainable resource may be applied if the rate of replacement exceeds the rate of use. The potential for being replaced in a relevant time frame is not itself sufficient to qualify a material as "renewable." The replacement or renewal must actually be occurring. There is still discussion on renewability issues and whether resource renewability as opposed to resource consumption belongs in inventory analyses or impact analyses. It is recommended that the inventory track and report each resource individually. Based on the lack of consensus among practitioners, at this time additional designations of natural resources into categories should be considered part of the impact analysis.

Energy

Energy as shown on the template (Figure 6) represents a combination of energy requirements for the subsystem. Three categories of energy are quantifiable—process, transportation, and energy of material resources (inherent energy).

Process energy is the energy required to operate and run the subsystem process(es), including such items as reactors, heat exchangers, stirrers, pumps, blowers, and boilers. Transportation energy is the energy required to power various modes of transportation such as trucks, rail carriers, barges, ocean vessels, and pipelines. Conveyors, forklifts, and other equipment that could be considered either transportation or process are labeled according to their role in the subsystem. For example, the power supplied to a conveyor used to carry material from one point in the subsystem to another point in the same subsystem would be labeled process energy. On the other hand, the power supplied to a conveyor used to transport material from one subsystem to a different subsystem would be considered transportation energy. Energy of material resources is described separately below.

Two alternatives exist for incorporating energy inputs in a subsystem module. One is to report the actual energy forms of the inputs, e.g., kilowatt-hours of electricity or cubic feet of natural gas. The other is to include the specific quantities of fuels used to generate the produced energy forms in the module.

The advantage of the first approach is that the specific energy mix is available for each subsystem. For example, a company may want to evaluate the desirability of installing a natural gas-fired boiler to produce steam compared to using its electrically heated boiler powered by a combination of purchased and on-site-generated electricity. A specific fuel mix could be applied to compute the energy and fuel resource use. The second approach, incorporating specific fuel quantities, allows a subsystem comparison of primary energy fuels. For example, "x" kWh of electricity would be specified as "y" ft^3 of natural gas and "z" lb of uranium.

Within each subsystem, the energy input data should be given as specific quantities of fuel and then converted into energy equivalents according to the conversion factors discussed in the following two sections. For example, the energy requirements attributed to a polyethylene resin plant may be specified as 500 lb of ethylene for feedstock, 500 ft^3 of natural gas, and 50 kWh of electricity to run the process equipment, and 50 gal of diesel fuel to transport the resin to consumers. In this case, the 50 kWh would be converted to 180 MJ.

Combustion and Precombustion Values. To report all energy usage associated with the subsystem of concern in the template, the analyst may need to consider energy data beyond the primary process associated with combustion of the fuel. The energy used in fuel combustion is only part of the total energy associated with the use of fuel. The amount of energy expended to acquire the fuel also may be significant in comparison to other energy expenditures. Energy to acquire fuel raw materials (e.g., mining coal or drilling for oil), process these raw materials into usable fuels, and transport them is termed by various practitioners as "precombustion energy" or "energy of fuel acquisition". Precombustion energy is defined as the total amount of energy necessary to deliver a usable fuel to the consumer of the fuel.

Including precombustion energy is analogous to extending the system boundaries for fuels to raw material inputs. For example, suppose the combustion of fuel oil in an industrial boiler results in the release of about 150,000 Btu per gallon. However, crude oil drilling and production, refining, and transporting the fuel oil require an additional 20,000 Btu per gallon. This additional energy is the precombustion energy. Thus, the total energy expended (precombustion energy plus combustion energy) when a gallon of fuel oil is consumed would be 170,000 Btu. Generally, a complete inventory will include precombustion energy contributions because they represent the true energy demand of the system. Inclusion or exclusion of this contribution should be clearly stated.

Energy Sources. Energy is obtained from a variety of sources, including coal, nuclear power, hydropower, natural gas, petroleum, wind, solar energy, solid waste, and wood biomass. Fuels are interchangeable, to a high degree, based on their energy content. For example, an electric utility decides which fuel or other energy source to use based on the cost per energy unit. Utilities can and do use multiple forms of energy sources, making possible an economic decision based on the energy cost per kilowatt-hour of electricity generated. Manufacturing companies also choose among energy sources on the same basis. However, reasons other than cost, such as scarcity or emissions to the environment, also affect the energy source decision. For example, during periods of petroleum shortages, finding products that use predominantly nonpetroleum energy sources may be desirable. For that reason, the inventory should characterize energy requirements according to basic sources of energy. Thus, it would consider not only electricity, but also the basic sources (such as coal, nuclear power, hydropower, natural gas, and petroleum) that produce the electricity.

Electricity. Considerations associated with electricity include the source of fuel used to generate the electricity and the efficiency of the generating system. Power utilities typically use coal, nuclear power, hydropower, natural gas, or oil to generate electricity. Non-utility generation sources can include wind power, waste-to-energy, and geothermal energy. Accurately determining electrical energy use and associated emissions raises several complications, such as relating the actual electricity use of a single user to the actual fuel used.

Although a given company pays its bills to a particular utility, the company is not simply purchasing power from the nearest plant. Once electricity is generated and fed into power lines, it is indistinguishable from electricity from any other source. Individual generating stations owned by a given utility may use different fuels. The electricity generated by these stations is "mixed" in the transmission lines of that utility. The utility is interconnected with neighboring utilities (also using various types of fuel), to form regional grids, which then interconnect to form a national grid.

Computational models currently used to perform life-cycle inventories of electricity in the USA are based on the fuel mix in regional grids or on a national average. In many cases where an industry is scattered throughout the USA, the fuel mix for the national grid (available from the U.S. Department of Energy) can be used, making calculations easier without sacrificing accuracy. Data for 1991 are shown in Table 1.

Table 1. U.S. National Electrical Grid Fuel Mix for 1991[a,b]

Fuel	Gigawatthours (GWh)	Percent
Coal	1,553,581	54.5
Nuclear	616,759	21.7
Hydro	291,657	10.2
Natural Gas	264,478	9.3
Oil	112,146	3.9
Other[c]	10,339	0.4
Total	2,848,960	100.0

(a) Sources: U.S. Department of Energy, Energy Information Agency, 1992; Canadian Electric Utilities and Natural Energy Board, 1991.

(b) Canadian data are estimates for 1991 based on 1990 generation and export. Canadian exports are 0.9% of totals and were equally allocated across fuel types.

(c) Includes wood and waste-to-energy sources but excludes independent generators and minor sources (e.g., geothermal).

One exception to the national grid assumption is the electroprocess industries, which use vast amounts of electricity. Aluminum smelting is the primary example. It and the other electroprocessing industries are not distributed nationally, so a national electricity grid does not give a reasonable approximation of their electricity use. They are usually located in regions of inexpensive electric power. Some plants have purchased their own electric utilities. In recognition of this fact, specific regional grids or data from on-site facilities are commonly used for life-cycle inventories of the electroprocessing industries.

The energy inefficiency of the electricity-generating and delivery system must also be considered. The theoretical conversion from the common energy unit of kilowatt-hours (kWh) to common fuel units (Joules) is 3.61 MJ per kWh. Ideally, the analyst would compute a specific efficiency based on the electrical generation fuel mix actually used. This value is derived by comparing the actual fuels consumed by the electricity-generating industry in the appropriate regional or national grid to the actual kilowatt-hours of electricity delivered for useful work. The value includes boiler inefficiencies and transmission line losses. However, a conversion of 11.3 MJ per kWh may be used in most cases to reflect the actual use of fuel to deliver electricity to the consumer from the national grid.

Nuclear Power. Nuclear power substitutes for fossil fuels in the generation of electricity. There is no measurement of nuclear power directly equivalent to the joules of fossil fuel, so nuclear power typically is measured at its fossil fuel equivalency. The precombustion energy of nuclear power is usually added to the fuel equivalency value. The precombustion energy includes that for mining and processing, as well as the increased energy requirements for power plant shielding.

Hydropower. Most researchers traditionally have counted hydropower at its theoretical energy equivalence of 3.61 MJ per kWh, with no precombustion impacts included. No precombustion factors are used for hydropower because water does not have an inherent energy value from which line transmission losses, etc., can be subtracted. The contribution of the capital equipment is small in light of the amount of hydroelectric energy generated using the equipment. Disruption to ecosystems typically has not been considered in the inventory. However, quantitative inventory measures that may be suitable for characterizing related issues, such as habitat loss due to land use conversion, potentially could be included. Factors addressing areal damage, recovery time, and ecosystem function are under consideration for inclusion in the impact analysis.

Energy Content of Material Resources. Many materials analyzed in a life-cycle inventory have an energy content. From a thermodynamic perspective, it is important to ensure that energy balance is maintained in each subsystem. However beyond this, further distinctions can be drawn.

The energy of material resources, also known as fuel-related inherent energy or latent energy of materials, accounts for those products or materials that consume raw materials whose alternative use is as a fuel or energy resource (e.g., oil, natural gas, and coal). Some materials are made from energy resources and, as stated earlier, inherent energy is a measure of the energy implications of the decision to forego use of the resource as fuel. The primary example is plastics, which are made from petroleum and natural gas. Because the actual plastic materials contain energy resources, resulting in a reduction of the planet's finite reserves, an energy value is assigned to the plastic material in addition to the other types of basic energy forms associated with plastic production, such as process and transportation energy. This additional energy value is equivalent to the fossil fuel combustion value of petroleum and natural gas.

In assessing the energy content issue for other material resources, different alternatives have been used by various researchers. The first option involves asking whether or not a given material is viewed as an energy resource. For materials such as coal, natural gas, and petroleum, there is no question. However, in the case of wood, textiles, and biomass crops, the answer is not as clear-cut. In the USA, wood generally is not viewed as a primary energy resource, so it typically has not been counted as such for U.S.-based studies. However, in performing studies in areas of the world where wood is a major energy resource, or for U.S. studies involving activities in these areas, it would be treated as a primary energy source.

The second option includes all raw materials having an energy content. This approach has been more widely but not universally used in Europe (Tillman, et al., 1991; Lundholm and Sundstrom, 1985). A third option, combining some features of the first two, is to track non-fuel inherent energy as a separate category.

Residues and Renewable Energy Sources. An important energy issue is the use of residues and energy sources from manufacturing operations, particularly those from agricultural or forest product operations. These sources of energy are listed commonly in the energy profile or characterization so that further analysis in the impact study can be made. However, the life-cycle inventory does not give special credit or benefit for the use of renewable energy resources. Renewable resource definition was discussed previously as a general raw materials issue.

Geographic Scope. Energy is an international product. All kinds of fuels are imported and exported, and electricity passes easily across national boundaries. Much of the crude oil used in the USA, for example, is produced in Middle Eastern countries such as Saudi Arabia. Historically, data on inputs and outputs associated with acquiring oil often have not been available for non-U.S. sources. Where data related to procurement of energy from a foreign source are not available, the approach has been to apply U.S. data as an initial estimate. When this approach is used, the analyst should clearly state the assumptions.

Energy from Waste Combustion. When waste is burned, energy can be recovered. The question is how to properly include the energy in an inventory. The energy value of combustible waste, whether industrial or postconsumer, historically is counted as the higher heating value (HHV) of the materials in the combusted waste with proper adjustment for moisture, just as fossil fuels are counted. However, there is no theoretical reason why the actual thermal yield of a waste of known composition cannot be determined. To calculate the thermal yield would require offsetting the HHV with both the moisture factor loss and the incinerator losses. Historically, the energy yield reduction associated with noncombustible materials introduced into an incinerator also has not been debited. Energy must be supplied to heat up these materials to the incinerator operating temperature. For postconsumer waste, proper accounting of the percentage sent to waste-to-energy facilities needs to be made. This energy value is then credited against the system energy requirements for the primary product, resulting in net energy requirements that are less than the total energy requirements for the system.

Water. Water volume requirements should be included in a life-cycle inventory analysis. In some parts of the country, water is plentiful. Along the coasts, seawater is usable for cooling or other manufacturing purposes. However, in other places water is in short supply and must be allocated for specific uses. Some parts of the country have abundant water in some years and limited supplies in other years. Some industrial applications reuse water with little new or makeup water required. In other applications, however, tremendous amounts of new water inputs are required.

How should water be incorporated in an inventory? The goal of the inventory is to measure, per unit of product, the gallons of water required that represent water unavailable for beneficial uses (such as navigation, aquatic habitat, and drinking water). Water withdrawn from a stream, used in a process, treated, and replaced in essentially the same quality and in the same location should not be included in the water-use inventory data. Ideally, water withdrawn from groundwater and subsequently discharged to a surface water body should be included, because the groundwater is not replaced to maintain its beneficial purposes. Data to make this distinction may be difficult to obtain in a generic study where site-specific information is not available.

In practice, the water quantity to be estimated is net consumptive usage. Consumptive usage as a life-cycle inventory input is the fraction of total water withdrawal from surface or groundwater sources that either is incorporated into the product, coproducts (if any), or wastes, or is evaporated. As in the general case of renewable versus nonrenewable resources, valuation of the degree to which the water is or is not replenishable is best left to the impact analysis.

Outputs of the Product Life-Cycle Inventory Analysis

A traditional inventory quantifies three categories of environmental releases or emissions: atmospheric emissions, waterborne waste, and solid waste. Products and coproducts also are quantified. Each of these areas is discussed in more detail in the following sections. Most inventories consider environmental releases to be actual discharges (after control devices) of pollutants or other materials from a process or operation under evaluation. Inventory practice historically has included only regulated emissions for each process because of data availability limitations. It is recommended that analysts collect and report all available data in the detailed tabulation of subsystem outputs. In a study not intended for product comparisons, all of these pollutants should be included in the summary presentations.

A comparative study offers two options. The first is to include in the summary presentation only data available for all alternatives under consideration. The advantage of this option is that it gives a comparable presentation of the loadings from all the alternatives. The disadvantage is that potentially consequential information, which is available only for some of the alternatives, may not be used. The second option is to report all data whether uniformly available or not. In using this option, the analyst should caution the user not to draw any conclusions about relative effects for pollutants where comparable data are not available. "Comparable" is used here to mean the same pollutant. For example, in a summary of data on a bleached paper versus plastic packaging alternatives, data on dioxin emissions may be available only for the paper product. The second option is recommended for internal studies and for external studies where proper context can be provided. A discussion of which pollutants are associated with various regulations is included in the appendix.

Atmospheric Emissions

Atmospheric emissions are reported in units of weight and include all substances classified as pollutants per unit weight of product output. These emissions generally have included only those substances required by regulatory agencies to be monitored but should be expanded where feasible. The amounts reported represent actual discharges into the atmosphere after passing through existing emission control devices. Some emissions, such as fugitive emissions from valves or storage areas, may not pass through control devices before release to the environment. Atmospheric emissions from the production and combustion of fuel for process or transportation energy (fuel-related emissions), as well as the process emissions, are included in the life-cycle inventory.

Typical atmospheric emissions are particulates, nitrogen oxides, volatile organic compounds (VOCs), sulfur oxides, carbon monoxide, aldehydes, ammonia, and lead. This list is neither all-inclusive nor is it a standard listing of which emissions should be included in the life-cycle inventory. Recommended practice is to obtain and report emissions data in the most speciated form possible. Some air emissions, such as particulates and VOCs, are composites of multiple materials whose specific makeup can vary from process to process. All emissions for which there are obtainable data should be included in the inventory. Therefore, the specific emissions reported for any system, subsystem, or process will vary depending on the range of regulated and nonregulated chemicals.

Certain materials, such as carbon dioxide and water vapor losses due to evaporation (neither of which is a regulated atmospheric emission for most processes), have not been included in most inventory studies in the past. Regulations for carbon dioxide are changing as the debate surrounding the greenhouse effect and global climate change continues and the models used for its prediction are modified. Inclusion of these emerging emissions of concern is recommended.

Waterborne Wastes

Waterborne wastes are reported in units of weight and include all substances generally regarded as pollutants per unit of product output. These wastes typically have included only those items required by regulatory agencies, but the list should be expanded as data are available. The effluent values include those amounts still present in the waste stream after wastewater treatment and represent actual discharges into receiving waters. For some releases, such as spills directly into receiving waters, treatment devices do not play a role in what is reported. For some materials, such as brine water extracted with crude oil and reinjected into the formation, current regulations do not define such materials as waterborne wastes, although they may be considered in solid waste regulations under the Resource Conservation and Recovery Act (RCRA). Other liquid wastes may also be deep well injected and should be included. In general, the broader definition of emissions in a life-cycle inventory, in contrast to regulations, would favor inclusion of such streams. It can be argued, from a systems analysis standpoint, that materials such as brine should count as releases from the subsystem because they cross the subsystem boundary. If wastes and spills that occur are discharged to the ocean or some other body of water, these values are always reported as wastes.

As with atmospheric wastes, waterborne wastes from the production and combustion of fuels (fuel-related emissions), as well as process emissions, are included in the life-cycle inventory.

Some of the most commonly reported waterborne wastes are biological oxygen demand (BOD), chemical oxygen demand (COD), suspended solids, dissolved solids, oil and grease, sulfides, iron, chromium, tin, metal ions, cyanide, fluorides, phenol, phosphates, and ammonia. Again, this listing of emissions is not meant to be a standard for what should be included in an inventory. Some waterborne wastes, such as BOD and COD, consist of multiple materials whose composition can vary from process to process. Actual waterborne wastes will vary for each system depending on the range of regulated and nonregulated chemicals.

Solid Waste

Solid waste includes all solid material that is disposed from all sources within the system. The regulations include certain liquids and gases in the definition as well. Solid wastes typically are reported by weight.

Some analysts convert the weight to volume using representative landfill density factors. By estimating landfill volumes, researchers can report the space occupied by the waste in a landfill. If volume factors are used, both weight and volume should be reported to enhance comparability among studies.

Distinction is made in data summaries between industrial solid wastes and post-consumer solid wastes, as they are generally disposed of in different ways and, in some cases, at different facilities. *Industrial solid waste* refers to the solid waste generated during the production of a product and its packaging and is typically divided into two categories: process solid waste and fuel-related solid waste. *Postconsumer solid waste* refers to the product/packaging once it has met its intended use and is discarded into the municipal solid waste stream.

Process solid waste is the waste generated in the actual process, such as trim or waste materials that are not recycled as well as sludges and solids from emissions control devices. Fuel-related waste is solid waste produced from the production and combustion of fuels for transportation and operating the process. Fuel combustion residues, mineral extraction wastes, and solids from utility air control devices are examples of fuel-related wastes.

Mine tailings and overburden generally are not regulated as solid waste. However, the regulations require overburden to be replaced in the general area from which it was removed. Furthermore, environmental consequences associated with the removal of mine tailings and overburden should be included. The regulations do not require industrial solid waste to be handled off site. Therefore, researchers try to report all solid waste from industrial processes destined for disposal, whether off site or local. Historically, no distinctions have been made between hazardous and nonhazardous solid waste, nor have individual wastes been specifically characterized. However, in view of the potentially different environmental effects, analysts will find it useful to account for these wastes separately, especially if an impact analysis is to be conducted.

Products

The products, as identified in the template, are defined by the subsystem and/or system under evaluation. In other words, each subsystem will have a resulting product, with respect to the entire system. This subsystem product may be considered either a raw material or intermediate material with respect to another system, or the finished product of the system.

Again using the bar soap system, when examining the meat packing subsystem, meat, tallow, hides, and blood would all be considered product outputs. However, because only tallow is used in the bar soap system, tallow is considered the only product from that subsystem. All other material outputs (not released as wastes or emissions) are considered coproducts. If the life-cycle assessment were performed on a product such as a leather purse, hides would be considered the product from the meat packing subsystem and all other outputs would be considered coproducts.

Although for bar soap the tallow is considered the product from the meat packing subsystem, it is simultaneously an intermediate material within the bar soap system. Thus, from these examples one can see that classifying a material as a product in a life-cycle study depends, in part, on the extent of the system being examined, i.e., the position from which the material is viewed or the analyst's point of view. This point of view should become clear when the template is applied to each subsystem within the total system under evaluation.

Transportation

The life-cycle inventory includes the energy requirements and emissions generated by the transportation requirements among subsystems for both distribution and disposal of wastes. Transportation data are reported in miles or kilometers shipped. This distance is then converted into units of ton-miles or tonne-kilometers, which is an expression involving the weight of the shipment and the distance shipped. Materials typically are transported by rail, truck, barge, pipeline, and ocean transport. The efficiency of each mode of transport is used to convert the units of ton-miles into fuel units (e.g., gallons of diesel fuel). The fuel units are then converted to energy units, and calculations are made to determine the emissions generated from the combustion of the fuels.

The template in Figure 8 shows that transportation is evaluated for the product leaving each subsystem. This method of evaluating transportation avoids any inadvertent double-counting of transportation energy or emissions. Transportation is reported only for the product of interest from a subsystem and not for any coproducts of the subsystem, because the destination of the coproducts is not an issue. The raw materials for the bar soap production system (Figure 7), for example, include salt from salt mining and trees from natural forest harvesting. Applying the template to these two subsystems shows that the transport of salt from the mining operation and the transport of trees from the logging operation must be included in the data collected for these subsystems. Logically, there is no transport of raw materials into these subsystems because the salt and trees were attached to the earth prior to removal.

The salt is transported to chlorine/sodium hydroxide plants, and the trees are transported to pulp mills. Applying the template to these subsystems shows that the transport of chlorine and sodium hydroxide from those plants to pulp mills is part of the chlorine production and sodium hydroxide subsystems. Likewise, the transport of pulp to paper mills is part of the pulp mill subsystem. The transport of raw materials, salt, and trees into the subsystems (chlorine production, sodium hydroxide production, and pulp mills) now being evaluated has already been accounted for in the evaluation of the salt mining and natural forest harvesting subsystems. Applying the template throughout the bar soap system shows the evaluation of transportation ending with the postconsumer waste management subsystem, where wastes may be transported to a final disposal site.

Backhauling may be a situation where there is some overlap between the transportation associated with product distribution and the transportation associated with recycling of the product or a different product after consumer use. A backhaul has been described as occurring when a truck or rail carrier has a profitable load in one direction and is willing to accept a reduced rate for a move in the return direction. Backhaul opportunities occur when the demand for freight transportation in one area is relatively low

Actual product flow diagram for the production of Products 'A' and 'B'

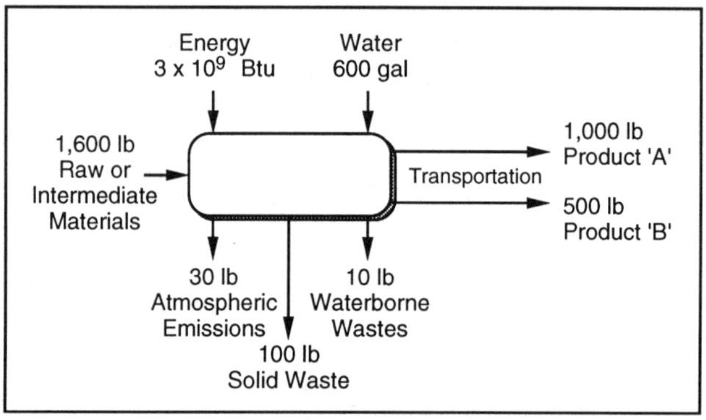

Flow diagrams showing the resources and environmental releases allocated to the two products.

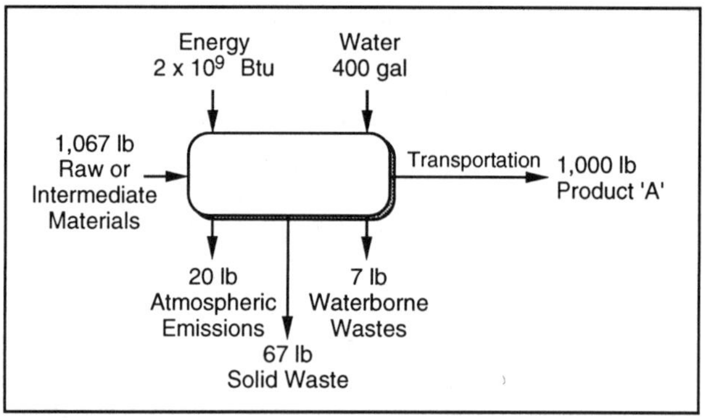

Coproduct Allocation for Product 'A'

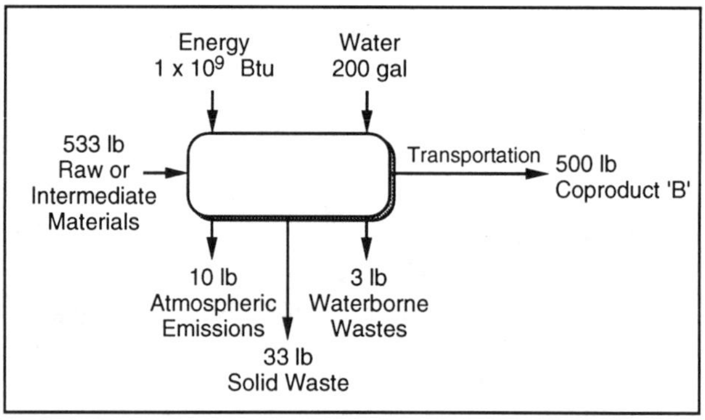

Coproduct Allocation for Product 'B'

Figure 8
Allocating resources and environmental burdens for a product and coproduct

Source: Franklin Associates, Ltd.

and carriers have a financial incentive to move their vehicles, loaded or empty, to a place where the demand for freight transportation is higher. Due to the lowered transportation rates, recycled materials, especially paper and aluminum, are often transported by backhauling. Thus, a carrier may take a load of new paper from a mill to customers in a metropolitan area and pick up loads of scrap paper in the same area to bring them back to the mill. In this scenario, backhauling may reduce the energy and emissions associated with distribution of a product by transferring the energy and emissions that would be associated with the empty return trip to the recycling stage.

Coproduct Allocation

Most industrial processes are physical and/ or chemical processes. The fundamentals of life-cycle inventory are based on modeling a system in such a way that calculated values reasonably represent actual (measurable) occurrences. Some processes generate multiple output streams in addition to waste streams. Usually, only certain of these output streams are of interest with respect to the primary product being evaluated. The term *coproduct* is used to define all output streams other than the primary product that are not waste streams and that are not used as raw materials elsewhere in the system examined in the inventory. A basis for coproduct allocation needs to be selected with careful attention paid to the specific items calculated. Figure 8 illustrates a common coproduct allocation scheme based on mass, the most common allocation basis used. However, each industrial system must be handled on a case-by-case basis. No allocation basis exists that is always applicable. Coproducts are of interest only to the point where they no longer affect the primary product. Subsequent refining of coproducts is beyond the scope of the analysis, as is transport of coproducts to facilities for further refining. In effect, the boundary for the analysis is drawn between the primary product and coproducts, with all materials and environmental loadings attributed to coproducts being outside the scope of the analysis. For example, the production of fatty acids from tallow for soap manufacture generates glycerine, a secondary stream that is collected and sold. Glycerine, therefore, is considered a coproduct, and its processing and use would be outside the scope of the bar soap analysis.

Basis for Coproduct Allocation

The first step is to investigate any complex process in detail and attempt to identify unit subprocesses that produce the product of interest. If sufficient detail can be found, no coproduct allocation will be necessary. The series of subprocesses that produce the product can simply be summed. Many metal manufacturing plants illustrate this approach. In steel product manufacture, all products are made by melting the raw materials, producing iron, and then producing raw steel. These steps are followed by a series of finishing operations that are unique to each product line. It is generally possible to identify the particular subprocesses in the finishing sequence of each product and to collect sufficient data to carry out the life-cycle inventory without coproduct allocation. In many cases, a careful analysis of unit systems will avoid the need to make coproduct allocations. Still, in some cases, such as a single chemical reaction vessel that produces several different products, there is no analytical method for cleanly separating the subprocesses. In this example, coproduct allocation is necessary.

The analyst needs to determine the specific resource and environmental categories requiring study. For a given product, different coproduct allocations may be made for different resource and environmental categories. To find the raw materials needed to produce a product, a simple mass balance will help track the various input materials into the output materials. For instance, if a certain amount of wood is needed to produce several paper products and the analysis concerns only one of the products, a mass allocation scheme as shown in Figure 8 will be used to determine the amount of wood required for the target product.

If a process produces several different chemical products, care must be taken in the analysis. It will be necessary to write balanced chemical equations and trace the chemical stoichiometry from the raw materials into the products. A simple mass allocation method frequently gives reasonable results, but not always. In calculating energy, heat of reaction may be the appropriate basis for allocating energy to the various coproducts. These calculations can become quite complex, but if the chemical products being produced are similar, experience has shown that a simple mass allocation very closely approximates the results of even more complex calculations.

If the various coproduct chemicals are quite different in nature, some other allocation method may be needed. For example, an electrolytic cell can produce hydrogen and oxygen from water. Each water molecule requires 2 electrons to produce 2 hydrogen atoms and 1 oxygen atom. On a macroscopic basis, electricity that produces 1 mole (or 2 grams) of hydrogen only produces one-half mole (or 16 grams) of oxygen. Thus, the input electrical energy would be allocated between the hydrogen and oxygen coproducts on a molar basis. That is, two-thirds of the energy would be allocated to the hydrogen and one-third to the oxygen, resulting in an energy per unit mass for hydrogen that is 16 times that of oxygen. However, conservation of mass is used to determine the materials requirements. Each mole of water (18 grams) contains 2 grams of hydrogen atoms and 16 grams of oxygen atoms, and the dissociation of the water results in 2 grams of hydrogen and 16 grams of oxygen. Thus, a mass allocation would be appropriate for raw material calculations in this example.

For environmental emissions from a multiple-product process, allocation to different coproducts may not be possible. For example, in a brine cell that produces sodium, chlorine, and hydrogen as coproducts, it may be tempting to associate any emissions containing chlorine with the chlorine coproduct alone. However, because the sodium and hydrogen are also produced by the same cell and cannot be produced from this cell without also producing chlorine, all emissions should be considered as joint wastes. The question arises as to how to allocate chlorine emissions (as well as other emissions) to all three products. There is no entirely satisfactory solution, but common practice based on chemical engineering, chemistry, and physics experience is to apply a mass allocation scheme as a reasonable modeling technique.

It has been suggested that the selling price of the coproducts could be used as a basis for this allocation. This solution is not entirely satisfactory because the selling prices of the various coproducts vary greatly with time and with independent competitive markets for each coproduct. Although a mass allocation basis may not be ideal, it is a widely recognized practice and produces a predictable and stable result.

It is necessary to carefully analyze each process and determine a basis for coproduct allocation based on the physical and chemical processes, and based on the resource and environmental parameters under study. However, when no such system can be agreed upon, a simple mass allocation can be used.

One final issue is the distinction between marginal wastes and coproducts. In some cases it is not clear whether a material is a waste or a coproduct. A hypothetical example might be a valuable mineral that occurs as 0.1% of an ore. For each pound of mineral product, 999 lb of unneeded material is produced. This discarded material might find use as a road aggregate. As such, it has value and displaces other commercial aggregates and appears to be a coproduct along with the valuable mineral. However, its value is so low that in some cases it may simply be dumped back on the ground because of limited markets. Whether this material is considered a waste or a coproduct may have a significant effect on the results of a product life-cycle inventory. It does not seem reasonable to use a simple mass allocation scheme here. It is more reasonable to assume that all of the energy and other resources and emissions associated with this process are incurred because of the desire for the valuable product mineral. However, there are some cases where the "waste" has marginal, but greater value than the example used here. It becomes difficult in these instances to determine precisely which of the coproduct allocation methodologies discussed above is most "correct".

One important role of an inventory is to provide information upon which impact analysis and improvement analysis can be based. In cases where there is no clear methodological solution, the inventory should include reasonable alternative calculations. It remains at some later time to make the judgments as to which of several reasonable alternatives is the correct one. In any event, it should be made clear what assumptions were made and what procedures were used.

Industrial Scrap

One coproduct stream of particular interest is *industrial scrap*. This term is used to specifically identify process wastes of value (trim scraps and off-spec materials) that are produced as an integral part of a manufacturing process. Further, the wastes have been collected and used as input materials for additional manufacturing processes. The last criterion is that these scrap materials have never been used as originally intended when manufactured. For example, a common polyurethane foam product is seat cushions for automobiles. The trim from cutting the cushions is never incorporated into seat cushions. Likewise, off-spec seat cushions sold as industrial scrap are never used as seat cushions, but are used as input material for another process.

A careful distinction must be made between industrial scrap and postconsumer waste for proper allocation in the inventory. If the industrial scrap is to be collected and used as a material input to a production system or process, it is credited in the life-cycle inventory as a coproduct at the point where it was produced. Unfortunately, systems that use material more efficiently, i.e., that produce lesser amounts of salable coproducts, assume a higher percentage of the upstream energy and releases using this criterion.

When the consumption of a coproduct falls within the boundaries of the analysis, it must no longer be considered as a coproduct, but as a primary product carrying with it all the energy requirements and environmental releases involved with producing it, beginning with raw materials acquisition. For example, a study of carpet underlayment made from polyurethane scrap would include the manufacturing steps for producing the polyurethane scrap. Its production must be handled as is any other subsystem of a life-cycle inventory. Industrial scrap does not displace virgin raw materials, because the consumption of the industrial scrap redefines the system to include the virgin materials for its production (isocyanates and polyalcohols in the case of polyurethane foam). Tallow (Figure 7) is another example of a material that would be defined as an industrial scrap/coproduct. Historically, the thinking has been that once a material shifts from the waste category to being a utilized material, or a coproduct, then it should bear some of the burden (energy, raw/intermediate material input, and environmental releases) for its own production.

DATA ISSUES

The gathering of data for each subsystem of the inventory is one of the greatest challenges in the inventory process. The defined purpose, goals, and objectives for the inventory will in part determine how these data issues are viewed or considered.

Data Quality

The quality of data used will significantly influence results. Because of the importance of data quality in life-cycle assessments, EPA is specifically addressing this issue in a separate guidelines document. This section presents an overview of some of the major data quality issues with respect to inventories.

Because life-cycle inventories are data-intensive and data quality can affect the outcome of an assessment, the development of uniform criteria is crucial for selection and reporting of data sources and types. Some basic objectives for data quality should be specified by the analyst based on the defined purpose for the study. From a data perspective, life-cycle inventories can be thought of as comprising two parts: a set of process and activity measurements that are amenable to standard statistical treatments, and a set of assumptions and decision rules for combining the data sets into a system. This discussion of data issues applies primarily to process and activity measurements. Considerations should include age of the data (because the technology on which data are based can become obsolete), frequency of data collection (ensuring that seasonal or other variability in the system is properly captured), and representativeness of the data (inclusion of the mix of activities that may contribute to an environmental loading). In addition, more traditional indices of data quality (accuracy, precision, detection limits, and completeness) should be evaluated with regard to life-cycle inventory applications.

The most accurate and recent data are desirable for performing a life-cycle inventory. All data received must be critically reviewed regarding the source and content before the data are used. Much of the data gathered for the performance of a life-cycle inventory is actual industry data, either direct facility measurements or indirect estimates from published summaries. Thus, accuracy is determined by the quality of the measurement or estimation procedures where all of the variables may not be known or controlled, and by the averaging process to obtain representative subsystems. Many plants produce several products, often from the same processes; thus some engineering estimation is often involved in getting representative data for a particular product or process. Many plants may produce the same product, but the processes, energy usages, and environmental loads may differ among plants. Thus, when data are gathered and averaged, the resulting data may not be characteristic of any existing plant.

Data confidentiality may also affect accuracy. Much of the data in a life-cycle inventory are from industry, either directly or indirectly. Ideally, companies using inventories publicly would report all data. If the inventory is to be publicly scrutinized and traceable, the data generally are aggregated to a more general level than if the data were guaranteed to be confidential. However, use of a peer review process at the most specific and detailed data collection level can help ensure that minimal concessions to accuracy loss are made and that variability measures are provided.

Specific and detailed data sources for all steps of the life cycle of a product are not always available, and researchers must resort to more general, and perhaps less accurate data sources, such as textbooks, periodicals, and public databases, which lack the level of detail desired for all the steps. In particular, formal data quality criteria often are not included in databases. Lack of data can be overcome in several ways. If the process in question is similar to other processes for which data are available, comparisons and estimates can be made. Much of the time, processes for which data are unobtainable or of uncertain quality represent a small portion of the entire system to be analyzed. Performing and applying a sensitivity analysis helps identify the relative importance of a particular step and can determine the amount of work necessary to obtain data that meet the needs of the inventory. At a minimum, the analyst should identify uncertain data and, if possible, estimate the degree of uncertainty.

Sensitivity Analysis

As noted in Chapter Three, sensitivity analysis is a systematic procedure for estimating the effects of data uncertainties on the outcome of a computational model. Applying sensitivity analyses to a life-cycle inventory begins early during the establishment of the boundaries and continues throughout the remainder of the inventory. It is not possible to establish a priori the significance of any individual contribution to the final inventory data set. The true test of whether an element is significant to an inventory is the sensitivity of the final result to the element's inclusion or exclusion, although experience allows trends to be discerned.

Typically, a sensitivity analysis is conducted by evaluating the range of uncertainty in the input data and recalculating the model's output to see the effect. Thus, higher levels of uncertainty in strongly influential variables will be less acceptable if the objectives of the study are to be met. Rules have been formulated to determine how much and in which combinations the inputs should be varied. As a rule if an estimate of the true variability, such as a measured statistical variance, is known, it should form the basis for the high- and low-range uncertainty estimates. The 95% confidence bounds are generally used for this purpose. For many inputs, the variability may not be known or may reflect variations in the feedstock over the period of time for which the inventory is being prepared. It is usual, in these instances, to vary the input by a range around its expected mean value. This may be 5% for some variables and an order of magnitude or more for others.

Frequently, in setting the boundaries of the inventory, that is, in answering the question of how far back to go, such order-of-magnitude estimates may be used to decide which input values require specification to a higher level of accuracy and which may be left "as is." Consider a simple, linear system with four inputs and an output as follows:

$$[a_1A_1 + (1-a_1)A_2] + B + C = D$$

where D is the total emission of an air pollutant, and A, B, and C, are individual unit process contributions to the total overall amount of D. Each process may be further divided into subprocesses. The fractions a_1 and $1-a_1$ determine the relative contributions from subprocesses to process A.

In setting up the boundaries, an analyst may want to decide if a process should be included or if order-of-magnitude estimates are acceptable. By running the calculations with and without a given process or by setting upper and lower ranges for the contribution, it is possible to decide whether a variable has a large or small effect. The rules for deciding simultaneous input variability are not well defined. However, the probability of all parameters simultaneously being at one extreme of the uncertainty range is low. Therefore, it is recommended that the analyst evaluate single variables at extremes of their estimated uncertainty range and then consider simultaneous variability of only the few most critical variables. These variables will either most influence the outcome or have the greatest uncertainty. In choosing which parameters to vary and in what amounts, the analyst should bear in mind that sensitivity analysis is a descriptive procedure. That is, its purpose is to differentiate more important and less important inputs, not to quantify overall uncertainty in the system.

Two purposes of sensitivity analysis are served by including and reporting it in every study. First, the analyst has a mechanism for deciding whether to expend additional effort to improve a critical data value or better characterize a subsystem. Second, decisionmakers have a map of the study showing quantitative areas of uncertainty. In many cases, the physical nature of the system under consideration dictates which inputs should be evaluated as a group, because changes in one variable may determine the range of others.

Accuracy and Precision

Data on specific inputs and emissions are preferable in an inventory. For example, data from the manufacturing plants and processes directly related to the material, product, process, activity, or service under consideration should be used whenever possible. Actual data should be used rather than estimates or regulatory limits. Assumptions and conventions used to gather and report data should be consistent and equitable.

Data should be collected at as detailed a level as possible, which allows for a more detailed analysis and reporting, and all emissions should be recorded at the same time. If aggregation has taken place prior to obtaining the data for the inventory, precluding disaggregation, it should be so noted.

Data in existing life-cycle inventories are sometimes reported in an aggregated form, such as listing all VOCs together. However, individual chemicals in such groups can have very different environmental properties. An adequate interpretation of these chemicals in the context of a life-cycle inventory requires that chemicals be listed individually. If the available data do not allow individual listing, or confidentiality precludes individual listing, this condition must be noted. Even more importantly, the analyst should avoid the implication that this type of categorical grouping can be compared across materials or life-cycle stages.

The various waste streams are characterized by measuring the concentrations of chemicals or of conventional parameters, e.g., COD and BOD, and the analytical methods used are reported. The variability of the measured data has to be taken into account. This can be done by listing the range of concentrations, minimum and maximum values, or a computed statistical deviation from the mean. For small data sets where statistical treatment would reveal the individual data points, a more descriptive statement of variability, e.g., less than a factor of two, is acceptable if the data are proprietary. In a number of cases, actual data will not be available for the volume of some emissions from waste treatment units. Sometimes it is possible to estimate the partitioning of a compound among the different media (air, water, solid waste). When such estimates are made, assumptions and calculations have to be reported. The sensitivity of the life-cycle inventory output to changes in this estimation should be explained when uncertainty exists about the accuracy of the estimation.

Data Source Attribution

Currently, little consistency exists regarding the specificity of source data used in life-cycle inventories. The degree of specificity needed is highly dependent on the scope of the study. In general, internal studies will contain more site-specific data. However, the analyst should recognize the difference between the reporting of inventory data on a non-site-specific basis and subsequent impact analyses that may require additional specificity. Further, no requirements have existed regarding specification of the type of source data used in studies. Because selection of varying qualities of source data can materially affect the life-cycle inventory results, the life-cycle inventory should document in detail the data source, how it was measured or calculated, and its type (e.g., average, worst-case, best-case, case-specific), and should include a discussion of its limitations, variability, and impact on the study results.

Uncertainty

During the data collection process, some gaps in data will likely be encountered. Several situations may occur as a result:

- *Absence of data from one of several producers of a particular product or material.* Where a variety of technologies exist, it may be more appropriate to assume the missing data are equal to the quantity averaged over only the plants reporting, assign that value to the missing data, and then average the total (Fava et al., 1991). For example, if only three of five producers report the data, then only three data points will be averaged to represent the entire process. The degree of completeness should be reported in the results presentation.

- *Lack of consistency on how many constituents are being recorded* (i.e., not every entity may collect the same data on a process). The best approach is to document the omission by making explicit that data were missing, not that the value was zero (i.e., during reporting, if summations are used they should be footnoted to explain that actual values may be higher or lower due to missing data) (Fava et al., 1991).
- *Depending on conventions, data may be reported as nondetectable or as less than a certain value* (the detection limit). If nondetectable entries are used, the detection limits should be reported. If data are reported as less than a certain value, that value should be used. Total numbers should be footnoted to indicate that actual values may be lower.

Some analysts report that the margin of significant difference in particular data categories for a life-cycle inventory is ±10-25%. Sensitivity analyses examining the high and low ranges of data points and weight ratios support this assertion. However, independent verification of these claims has not been made. The nature of the overall accounting error is predominantly systematic, and is thus determined by data sources and methods rather than by measurement randomness. In consequence, alternatives where the solid waste volumes, for example, differ by only 6% would not be significantly different. The analyst should clearly communicate when the data do or do not support a conclusion about differences among alternatives.

Representativeness

To ensure representativeness, characteristics of the sample must correspond sufficiently closely to the studied population to represent that population. Some issues regarding representativeness can be resolved by reviewing the experimental design and the way results are reported. At a minimum, the degree to which the reported data encompass the population should be stated. Such a statement for the bar soap example might be, "This study uses data covering 83% of domestic tallow production."

Occasionally inventory data are difficult to obtain for certain steps of the system or even for certain inputs or outputs of a subsystem. It is important to evaluate critically whether it is worth the time and effort to obtain data for some of the minor subsystems of the study. When literally thousands of numbers are involved in an analysis, most individual numbers contribute little to the overall analysis, especially if they have to do with a relatively minor component. Sometimes data for similar processes can be used to estimate the data for a minor component of the system. If data are carefully estimated in this way, potential error from estimating the data can be minimized. The importance of any data input on the results can be determined by performing a sensitivity analysis.

Data Time Period

The time period that data represent should be long enough to smooth out any deviations or variations in the normal operations of a facility. These variations might include plant shutdowns for routine maintenance, startup activities, and fluctuations in levels of production. Often data are available for a fiscal year of production, which is usually a sufficient time period to cover such variations.

Specific Data versus Composite Data

When the purpose of the inventory is to find ways to improve internal operations, it is best to use data specific to the system that is being examined. These types of data are usually the most accurate and also the most helpful in analyzing potential improvements to the environmental profile of a system. However, private data typically are guarded by a confidentiality agreement, and must be protected from public use by some means. Composite, industry-average data are preferable when the inventory results are to be used for broad application across the industry, particularly in studies performed for public use. Although composite data may be less specific to a particular company, they are generally more representative of an industry as a whole. Such composite data can also be made publicly available, are more widely usable, and are more general in nature. Composite data can be generated from facility-specific data in a systematic fashion and validated using a peer review process. Variability, representativeness, and other data quality indicators can still be specified for composite data.

Geographic Specificity

Natural resource and environmental consequences occur at specific sites, but there are broader implications. It is important to define the scope of interest (regional vs. national vs. international) in an inventory. A local community may be more interested in direct consequences to itself than in global concerns.

In general, most inventories done domestically relate only to that country. However, if the analysis considers imported oil, the oil-field brines generated in the Middle East should be considered. It has been suggested that the results of life-cycle inventories indicate which energy requirements and environmental releases (of the total environmental profile of a product) are local. However, due to the fact that industries are not evenly distributed, this segmenting can be done only after an acceptable level of accuracy is agreed on. The United States, Canada, Western Europe, and Japan have the most accurate and most readily available information on resource use and environmental releases. Global aspects should be considered when performing a study on a system that includes foreign countries or products, or when the different geographic locations are a key difference among products or processes being compared. As a compromise, when no specific geographical data exist, practices that occur in other countries typically are assumed to be the same as for their domestic counterparts. These assumptions and the inherent limitations associated with their application should be documented within the inventory report. In view of the more stringent environmental regulations in developed countries, this assumption, while necessary, often is not correct. Energy use and other consequences associated with importing materials should also be included.

Technology Mixes/Energy Types

For inventory studies of processes using various technology mixes, market share distribution of the technologies may be necessary to accurately portray conditions for the industry as a whole. The same is true of energy sources. Most inventories can be based on data involving the fuel mix in the national grid for electricity. There are exceptions, such as the aluminum electroprocessing industry previously discussed. Variations of this kind must be taken into account when applying the life-cycle inventory methodology. Also, as previously mentioned, conditions can differ greatly across international borders.

Data Categories

Environmental emission databases usually cover only those items or pollutants required by regulatory agencies to be reported. For example, as previously mentioned, the question of whether to report only regulated emissions or all emissions is complicated by the difficulty in obtaining data for unregulated emissions. In some cases, emissions that are suspected health hazards may not be required to be reported by a regulatory agency because the process of adding them to the list is slow. A specific example of an unregulated emission is carbon dioxide, which is a greenhouse gas suspected as a primary agent in global warming. There is no current requirement for reporting carbon dioxide emissions, and it is difficult to obtain measured data on the amounts released from various processes. Thus, results for emissions reported in a life-cycle inventory may not be viewed as comprehensive, but they can cover a wide range of pollutants. As a rule it is recommended that data be obtained on as broad a range as possible. Calculated or qualitative information, although less desirable and less consistent with the quantitative nature of an inventory, may still be useful.

Routine/Fugitive/Accidental Releases

Whenever possible, routine, fugitive, and accidental emissions data should be considered in developing data for a subsystem. If data on fugitive and accidental emissions are not available, and quantitative estimates cannot be obtained, this deficiency should be noted in the report of the inventory results. Often estimates can be made for accidental emissions based on historical data pertaining to frequency and concentrations of accidental emissions experienced at a facility.

When deciding whether to include accidents, they should be divided into two categories, based on frequency. For the low-frequency and high-magnitude events, e.g., major oil spills, tools other than life-cycle inventory may be appropriate. Unusual circumstances are difficult to associate with a particular product or activity. More frequent, lower magnitude events should be included, with perhaps some justification for keeping their contribution separate from routine operations.

SPECIAL CASE BOUNDARY ISSUES

In all studies, boundary conditions limiting the scope must be established. The areas of capital equipment, personnel issues, and improper waste disposal typically are not included in inventory studies, because they have been shown to have little effect on the results. Earlier studies did consider them in the analysis; later studies have verified their minimal contribution to the total system profile. Thus, exclusion of contributions from capital equipment manufacture, for example, are not excluded a priori. The decision to include or not to include them should be clearly noted by the analyst.

Capital Equipment

The energy and resources that are required to construct buildings and to build process equipment should be considered. However, for most systems, capital expenditures are allocated to a large number of products manufactured during the lifetime of the equipment. Therefore, the resource use and environmental effluents produced are usually small when attributed to the system of interest. The energy and emissions involved with capital equipment can be excluded when the manufacture of the item itself accounts for a minor fraction of the total product output over the life of the equipment.

Personnel Issues

Inventory studies focus on the comprehensive results of product consumption, including manufacturing. At any given site, there are personnel-related effluents from the manufacturing process as well as wastes from lunchroom trash, energy use, air conditioning emissions, water pollution from sanitary facilities, and others. In addition, inputs and outputs during transportation of personnel from their residence to the workplace can be significant, depending on the purpose and scope of the inventory. In many situations, the personnel consequences are very small and would probably occur whether or not the product were manufactured. Therefore, exclusion from the inventory may be justified. The analyst should be explicit about including or excluding this category. For these issues, the goals of the study should be considered. If the study is comparative and one option is significantly different in personnel or capital equipment requirements, then at least a screening-level evaluation should be performed to support an inclusion or exclusion decision.

Improper Waste Disposal

For most studies it is assumed that wastes are properly disposed into the municipal solid waste stream or wastewater treatment system. Illegal dumping, littering, and other improper waste disposal methods typically are not considered in life-cycle inventories as a means of solid waste disposal. Where improper disposal is known to be used and where environmental effects are known or suspected, a case may be made to include these activities.

Chapter Five

ISSUES APPLICABLE TO SPECIFIC LIFE-CYCLE STAGES

INTRODUCTION

This chapter discusses the issues specific to each of the four life-cycle stages, i.e., raw materials and energy acquisition, manufacturing, use/reuse/maintenance, and recycle/waste management. In addition, the steps in manufacturing are discussed individually. The subsystem boundaries, as well as specific assumptions and conventions, are discussed for each stage and step.

Major Concepts—Raw Materials

- The resource requirements and environmental emissions are calculated for all of the processes involved in acquiring raw materials and energy. This analysis involves tracing materials and energy back to their sources.
- Consequences of the raw materials acquisition stage include nontraditional inventory outputs, such as land use changes, and nonchemical releases, such as odor or noise. To the extent they are quantifiable, such outputs may be incorporated.
- When fuel sources become input materials for a manufacturing process, an energy factor accounts for the unused energy inherent in the fuel.

RAW MATERIALS ACQUISITION STAGE

The life cycle of any product or material begins with the acquisition of raw materials and energy sources. For example, crude oil and natural gas must be extracted from drilled wells, and coal and uranium must be mined before these materials can be processed into usable fuels. All of these activities fall into the raw materials acquisition stage.

Subsystem Boundaries

The subsystem boundaries for raw materials acquisition encompass the actual process(es) of acquiring the raw material, i.e., obtaining the material from the earth or earth's surface as it naturally occurs. Raw materials acquisition includes any energy and water used and environmental releases. Other consequences that are measurable, such as land use changes, also may be included. Transport of the raw materials to the point of refining and processing is also included within the boundaries of the raw materials acquisition subsystems.

Figure 9 illustrates the manufacture and use of bar soap on a life-cycle basis. The shaded areas show the subsystems that are included in the raw materials acquisition stage of this system. In this example, the raw materials acquisition stage includes the

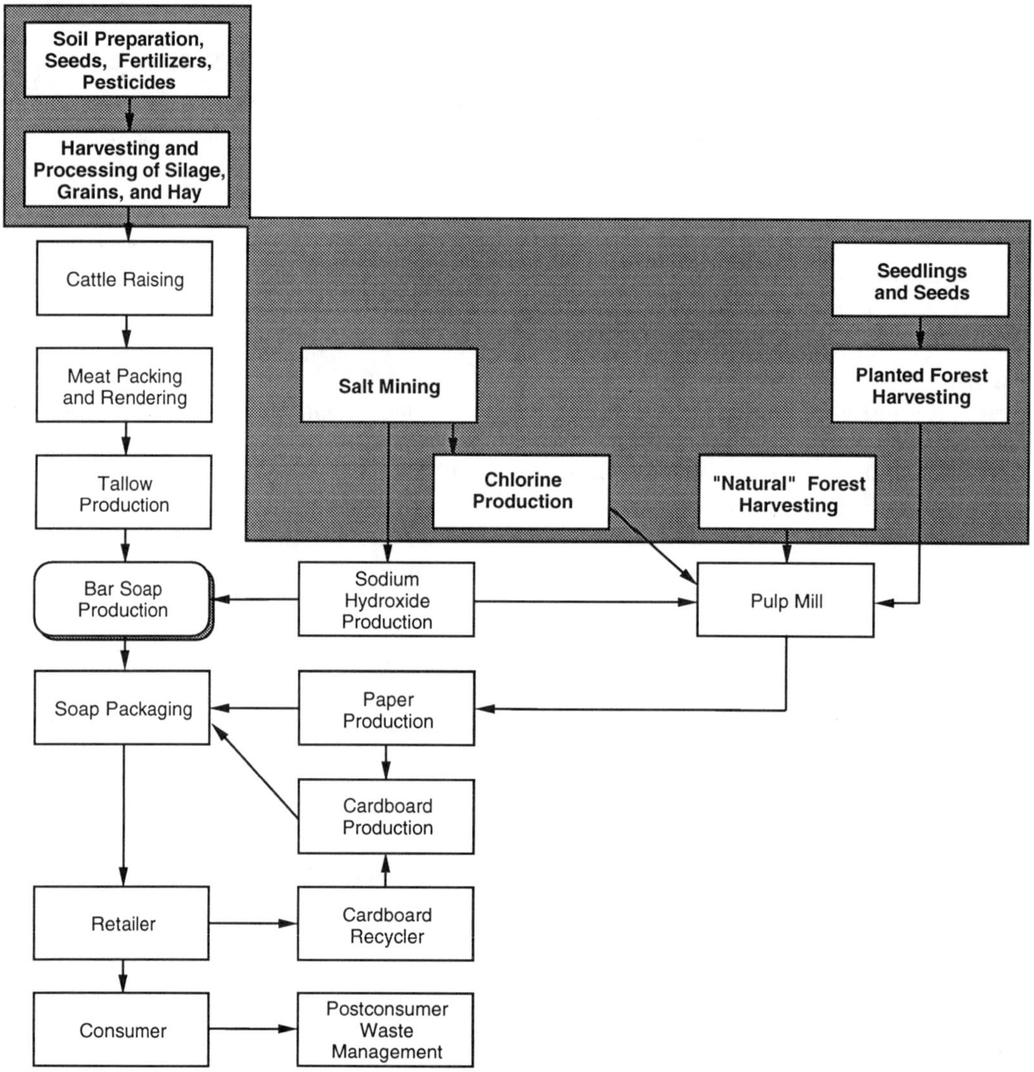

Note: Energy Fuels and Electric Generation Subsystems are inputs to all subsystems using on-site-generated or purchased electricity.

Figure 9
The first step in a product's life cycle is the acquisition of raw materials and energy

planting, harvesting, and processing of silage, grains, and hay; salt mining; natural forest harvesting; and the planting and harvesting of planted forests. The energy expended and water used in acquiring the raw materials is included in this stage. The environmental emissions and wastes produced in the activities related to raw materials acquisition also are included in this stage.

A typical life-cycle inventory of a product evaluates the primary product and may include associated primary, secondary, and tertiary packaging. For example, the analysis of bar soap includes the manufacture of tallow and sodium hydroxide for soap making, paper for packaging the soap (primary packaging), and corrugated boxes for shipping the soap (secondary packaging). The manufacture of each material included in the analysis begins with the acquisition of the raw materials necessary to produce each component. Any fuel, chemical, additive, or material that is significant enough to be included in the analysis is traced back to its raw material acquisition step. For example, large quantities of fertilizers and pesticides are used to produce silage, grain, and hay, which are the raw materials for tallow production. Thus, the energy and emissions for the production of the fertilizers and pesticides need to be calculated in the inventory, including the mining and crude oil production steps to acquire the raw materials used to produce the fertilizers and pesticides.

In a traditional inventory, a material contributing less than 1% by weight of the total system typically contributes less than 1% to the total emissions of a particular inventory item and can be omitted from the inventory. The "less than 1%" effect is generally considered insignificant. This approach is based on many years of conducting such studies and not on statistical or technical grounds. However, it does not make any assumptions regarding the environmental significance of the emission. Thus, one problem with a blanket application of this approach is that toxic materials could inadvertently be eliminated from the inventory analysis even though they present a serious potential environmental impact. Most practitioners examine the system and make exceptions to a strict interpretation of the "less than 1%" rule where toxic materials are concerned.

In practice, the significance of a specific item to the inventory may be determined by performing a sensitivity analysis of that item. If omission of the item does not affect the ability to use the data to support the purpose of the inventory, the item may be excluded from the study. For instance, the printing ink on a folding carton may be less than 1% by weight of the system and probably can be eliminated from the system with little effect on overall data accuracy. However, the printing ink on a thin plastic film may be 10% or more by weight of a different system and probably should be factored in to maintain the overall accuracy desired. If concerns about toxic constituents in the ink are an issue, two options are available. One is to include the ink in the folding carton inventory. When this is done, however, depending on the number of product units for which the final data are reported, the emissions associated with the ink can get lost in the overall data. The other option is to conduct a separate inventory on the ink.

Another example would be the cleaning agents or routine maintenance chemicals used in a process, such as volatile organic solvents used to clean a printing press. If the system being examined is large in scope, these chemicals will probably be shown by sensitivity analysis to be insignificant. If the study is focused on a specific process, such as the printing operation itself, the cleaning agents and maintenance chemicals probably will be significant. Therefore, if a material or chemical is a significant part of the system, the inventory should include its production, including the raw material acquisition steps for producing it.

Specific Assumptions and Conventions

Raw Materials Acquisition

Analysis of a material or product begins with specific data for the acquisition of raw materials. For example, analysis of the manufacture of the paper used to package a bar of soap will begin with the mining of salt and harvesting of trees, because these are the raw materials needed to produce paper. Specific assumptions for the acquisition of raw materials are listed below:

General 1. The acquisition of a raw material or energy source requires a disturbance to the environment. Ecosystems are impacted in the harvesting of trees, in mining for minerals, in using land to produce agricultural crops, and in drilling for petroleum or natural gas. Resource requirements that can be quantified should be included in the life-cycle inventory. Statistics are available that quantify such effects as pesticide runoff from agricultural activities, brine production from oil wells, waterborne wastes from animal feedlots, etc. However, other consequences of raw materials acquisition are not routinely measured and would be difficult to quantify, such as soil erosion, damage to watersheds, thermal pollution, and habitat destruction. If measures of these outputs can be provided by the analyst, they can be included in the inventory. There is presently no consensus among practitioners regarding the inclusion of qualitative factors in the text of an inventory report. These would be covered in an impact analysis.

General 2. Traditional fuels such as petroleum, natural gas, or coal are sometimes used as raw materials. For example, crude oil and natural gas are raw materials for plastic products. When these traditional fuels are used as raw materials, they are assigned an inherent energy value (also called energy of material resource or latent energy) equal to the heat of combustion of the raw material because the fuels have been removed from the total fuel supply.

Because wood is not commonly used as a fuel in the USA and the wood that is used as a fuel in industry is typically a waste product (e.g., bark from a pulp or paper mill), several options for handling the energy value of wood deserve consideration. First, it is unlikely that wood waste would be used as fuel for another industry within the USA because of its location, so it can be assumed that it would become a solid waste if it were not burned at the paper mill. Thus, one option is to avoid giving an inherent energy value to wood when it is used as a raw material in the USA. If the study were performed in a country where wood is a primary fuel, then an energy of material resource value equal to the wood's heat of combustion would be assigned to any wood used as a raw material.

A second option for handling the wood's energy value would be to count wood in a nonfuel category, especially in the USA. This option would require the addition of a separate category to record nonfuel inherent energy, as discussed in item 4 below.

A third option that has been proposed for considering the energy value of wood waste is to assume that wood burned at a paper mill to generate energy reduces the amount of traditional fuels that are burned. For this option, inherent energy credit would be given to this process for reducing the amount of traditional energy materials required by burning the wood. Because this option modifies the intent of a life-cycle inventory to quantify actual consequences, it is not recommended.

General 3. In a life-cycle inventory, energy requirements of a system are not reduced or credited for the use of "renewable" resources instead of "nonrenewable" resources. For example, a process may use wood energy instead of coal energy. The issues or distinctions between the levels of renewability are difficult to define and even more difficult to value in terms of energy. Furthermore, an inventory analysis is designed to quantify energy and raw materials requirements. The impact analysis is the appropriate place to characterize effects such as renewability.

General 4. To ensure that all sources of energy are properly accounted for in the inventory analysis, it may be necessary to create a separate category of nonfuel inherent energy that will permit closing the energy loop in a thermodynamic sense without artificially inflating the inherent energy values for fuels. This separate category for nonfuel inherent energy would not be reported in the summary tables of energy, but would still permit reviewers to account for all energy inputs and outputs. Examples of inherent energy materials that typically are not considered to be fuels include biomass waste materials such as sugar cane bagasse, wood residues from logging operations, and textile fibers.

Mining Operations 1. Overburden, the material overlying the ore or material being mined, is not considered to be an environmental emission (i.e., a solid waste) in mining operations. This is because the overburden is returned to the same land rather than being landfilled after the mining operations are completed. However, land use changes occur with this operation and may be quantified in the inventory. Habitat effects associated with mining may occur, but these are treated in the impact analysis, not in the inventory analysis. However, other quantifiable emissions such as particulates released into the air are included in the inventory.

Petroleum and Natural Gas Acquisition 1. Brine water is a coproduct of the production of crude oil. A portion of the brine water is reinjected into separate wells designed to receive it. The waterborne wastes contained in the part of the oil or gas well brine water that is reinjected are not included in the inventory because the wastes are essentially being returned to the same location. In an inventory, it is assumed that groundwater flow and quality are not affected. This assumption may be examined in an impact analysis. Waterborne wastes discharged into the ocean or another body of water are included.

Petroleum and Natural Gas Acquisition 2. Natural gas is often produced in combination with crude oil.Therefore, environmental emissions from drilling operations are allocated between crude oil and natural gas production. This allocation is made based on historical production data. For example, a drilling operation may produce 50% crude oil and 50% natural gas by weight. Thus, the data for the process will be split between them, using one of the coproduct allocation techniques discussed in Chapters Three and Four.

Lumbering Operations 1. Wood residue left in forests after tree harvesting is not considered to be a solid waste because it is not landfilled. In most logging operations, the residue is left to decompose where the lumber was harvested. However, in some operations the harvested land may be burned off, which generates atmospheric emissions that must be quantified. Any assumptions made as to waste practices in the harvest of wood must be thoroughly documented. Other outputs from lumbering operations that are quantifiable, such as land use changes, may be included.

Agricultural Raw Materials and Animal Products 1. Harvesting of agricultural products often involves significant manual labor. This is especially true in developing countries. Energy requirements and environmental emissions related to sustaining human life (e.g., producing food, clothing, or shelter) typically are not included in the life-cycle inventory.

Agricultural Raw Materials and Animal Products 2. In a life-cycle inventory analyzing the use of an animal product, the feed to produce the animal product usually is considered to be the main raw material for the system. For example, the analysis of tallow for bar soap may have corn as a raw material for the system. Corn is fed to the cattle that produce the tallow. Acquiring the raw material (corn) in this system requires energy for planting, harvesting, and transporting the corn. Emissions from pesticides and herbicides are associated with producing corn as a raw material. Therefore, the raw material acquisition steps for those chemicals should be included in the system.

Calculating Resource Requirements and Emissions for Raw Material Acquisition. Using the bar soap example (Figure 9), the raw material for sodium hydroxide is salt. The energy requirements and environmental emissions associated with the mining of salt and its transport to a caustic production facility are the energy and emissions of raw materials acquisition. Assume that the hypothetical data below represent the energy and emissions for mining 1,000 pounds of salt.

Process Energy

Electricity	50	kilowatt-hours
Coal	50	pounds
Residual oil	50	gallons

Transportation Energy

Ocean transport	50	ton-miles
Diesel	5	gallons

Atmospheric Emissions

Particulates	5	pounds

Assume further that 5 pounds of salt must be mined for every pound of sodium hydroxide manufactured and used. To determine the energy and emissions associated with raw materials acquisition for the manufacture of 1,000 pounds of sodium hydroxide, the above data would be multiplied by five in the computer spreadsheet.

Energy Acquisition

The energy requirements and environmental emissions attributed to the acquisition, transportation, and processing of fuels to a usable form are labeled as precombustion energy or emissions. Whenever a specific fuel is used in any of the processing or transportation steps, the appropriate quantities of precombustion energy and emissions are included in the total energy and emissions attributed to the use of that fuel. Therefore, when gasoline is used as a fuel, the energy and emissions must account not only for those from the combustion of the gasoline, but also for those attributable to the raw material extraction, refining, and processing required to produce the gasoline before it is burned. The assumptions listed for the acquisition of raw materials also generally apply to the acquisition of energy raw materials. The five basic energy sources included in a life-cycle analysis are coal, petroleum, natural gas, nuclear power, and hydropower. In certain cases, wood may also be included. Minor energy sources such as biomass, solar, and geothermal energy are generally ignored unless significant to a specific process.

Specific Steps for Calculating Energy Acquisition. Residual oil is one fuel used in several of the individual processes in the bar soap manufacturing system. Assume that every use of residual oil has been accounted for in this system and that the total used is 50 gallons. Assume that the hypothetical data listed below represent the total energy and emissions for producing 1 gallon of residual oil, including extracting crude oil, transporting it, and processing it into residual oil.

Process Energy

Electricity	10 kilowatt-hours
Coal	10 pounds
Natural gas	10 cubic feet

Transportation Energy

Pipeline	50 ton-miles
Natural gas	5 cubic feet

Atmospheric Emissions

Particulates	0.25 pound
Hydrocarbons	0.25 pound

Solid Wastes	0.25 pound

The hypothetical data above must be multiplied by 50 to calculate the precombustion energy and emissions associated with the 50 gallons of residual oil used in the bar soap system. These are the total energy requirements and emissions associated with the acquisition of 50 gallons of residual oil to be used as process energy in the system. A computer spreadsheet is used to make similar calculations for every other fuel used in the system.

Electrical Energy Acquisition

The production of electrical energy is more complex than simply refining a fuel. Therefore, this section addresses electrical energy acquisition separately. Utility power plants generate electricity from five basic energy sources: coal, petroleum, natural gas, nuclear power, and hydropower. A small percentage of electricity is also generated by unconventional sources such as biomass, solid waste, solar, and geothermal energy. The acquisition of electrical energy includes extracting the basic energy sources from the earth, processing these energy sources into usable fuels, and converting the fuels into electrical energy. As noted previously, the manufacturing processes for the production of any given product are sufficiently scattered throughout the country that the national average fuel mix for the electric utilities is representative when used for all of the manufacturing steps in an analysis. An exception to this assumption is the electroprocess industries, with aluminum smelting as the primary example. These industries locate in areas of inexpensive electrical power because they require vast amounts of electricity. The fuel mix for regional utility grids may be used in these specific situations. Historical data provide the basis for determining the efficiency of converting the different fuels to electricity. Historical data are also available for determining the transmission line losses (i.e., difference between amounts of electricity generated and sold) for delivery of electrical energy to the consumer.

Specific Steps for Calculating Electrical Energy Acquisition. Three steps are required to determine the total energy and emissions associated with the use of electrical power: extraction of fuel sources, processing of fuels, and converting the fuels to electrical energy. Before any of these steps may begin, the fuel mix for the electricity being used should be calculated. As expressed above, the national average fuel mix for the electric utilities is usually adequate for most industries. Where or when possible, electricity for a specific industry should be traced back to its specific fuel source to develop a more accurate profile for each process. For example, assume that the utilities are using 50% coal and 50% residual oil. Thus, the energy and emissions associated with coal mining and crude oil extraction, as well as transport of these materials, must be quantified. Any processing necessary to make these fuels usable must be incorporated into the analysis. After processing, the fuels are burned to produce steam for generating electrical energy. Both the emissions from the combustion of the fuels and the efficiency of the boiler system must be included in the analysis. Finally, the transmission line losses (the difference between the electricity generated and the electricity delivered) must be accounted for. All of these processes fall into the raw materials acquisition phase of the life-cycle inventory.

MANUFACTURING STAGE

The second stage of a product's life cycle is manufacturing. Manufacturing results in the transformation of raw materials into products delivered to end users. The manufacturing stage is further divided into three steps: materials manufacture, product fabrication, and filling/packaging/distribution. Each of these is discussed in the following sections.

Major Concepts—Materials Manufacture

- Materials manufacture converts raw materials into the intermediate products from which the finished product will be fabricated.
- Material scrap from a subsystem can be reused internally, sold as industrial scrap, or disposed of as solid waste. The inventory account for each option is handled differently.
- No credits or debits are applied to the subsystem for internally recycled material because no material crosses a subsystem boundary.
- Industrial scrap as a coproduct carries with it the energy and wastes to produce it. This ensures consistency with operations that use the scrap in house.

Materials Manufacture Step

The first step in manufacturing is the manufacture of materials. This step includes all manufacturing processes required to process raw materials into the intermediate materials from which the finished product will be fabricated. In an inventory examining bar soap, this step would include all operations required to produce tallow and sodium hydroxide from which bar soap is made. Similarly for paper packaging, this step would include all operations required to transform wood into paper.

Subsystem Boundaries

The boundaries for material manufacture subsystems encompass the actual process(es) of manufacturing an intermediate material, either from raw materials or from other intermediate materials. This step includes any energy, material, and water input and the environmental releases. Transport of the intermediate material produced to the site of the next manufacturing process or to the point of product fabrication is also included in the boundaries of a material manufacturing subsystem. Depending on the locations of facilities where material manufacture occurs, nontraditional inventory outputs such as land use and odors may be relevant. Any number of material manufacture subsystems may be required to convert a raw material into the intermediate material required to fabricate a product. For example, to convert crude oil into a polyethylene milk bottle, three subsystems are required: crude oil refining, ethylene production, and polyethylene production. Each subsystem has its own boundaries as described above to facilitate data gathering and to eliminate double-counting or omissions. The boundaries of the material manufacture step are illustrated for bar soap by the shading in Figure 10. Each of the material manufacturing operations can be viewed as a subsystem within the product system. Data for each manufacturing operation or subsystem are gathered separately. Thus, the material manufacturing step is not necessarily linear, but may be a complex arrangement of processes and multiple raw materials.

For each subsystem, the materials and energy inputs for processing are analyzed. The air and water emissions and the solid wastes resulting from each subsystem are also reported. In other words, a material and energy balance must be performed on each operation within the system. Energy and environmental wastes resulting from the transportation required from one process operation to another, or to the point of product fabrication, are also included.

Specific Assumptions and Conventions

The following assumptions and conventions generally apply to this step:

Coproduct Allocation. Manufacturing processes, particularly chemical processes, often produce more than one product. In most cases, only one of those products is used in the system being evaluated. Thus, some allocation of material and energy inputs and waste and emission outputs is necessary for marketed output materials. Chapter Four presents several methods for allocating requirements and emissions to various coproducts.

Industrial Scrap. Many manufacturing processes generate scrap material. This material is often reused within the same manufacturing process, reducing the new material input into the system. In such a case, the system is

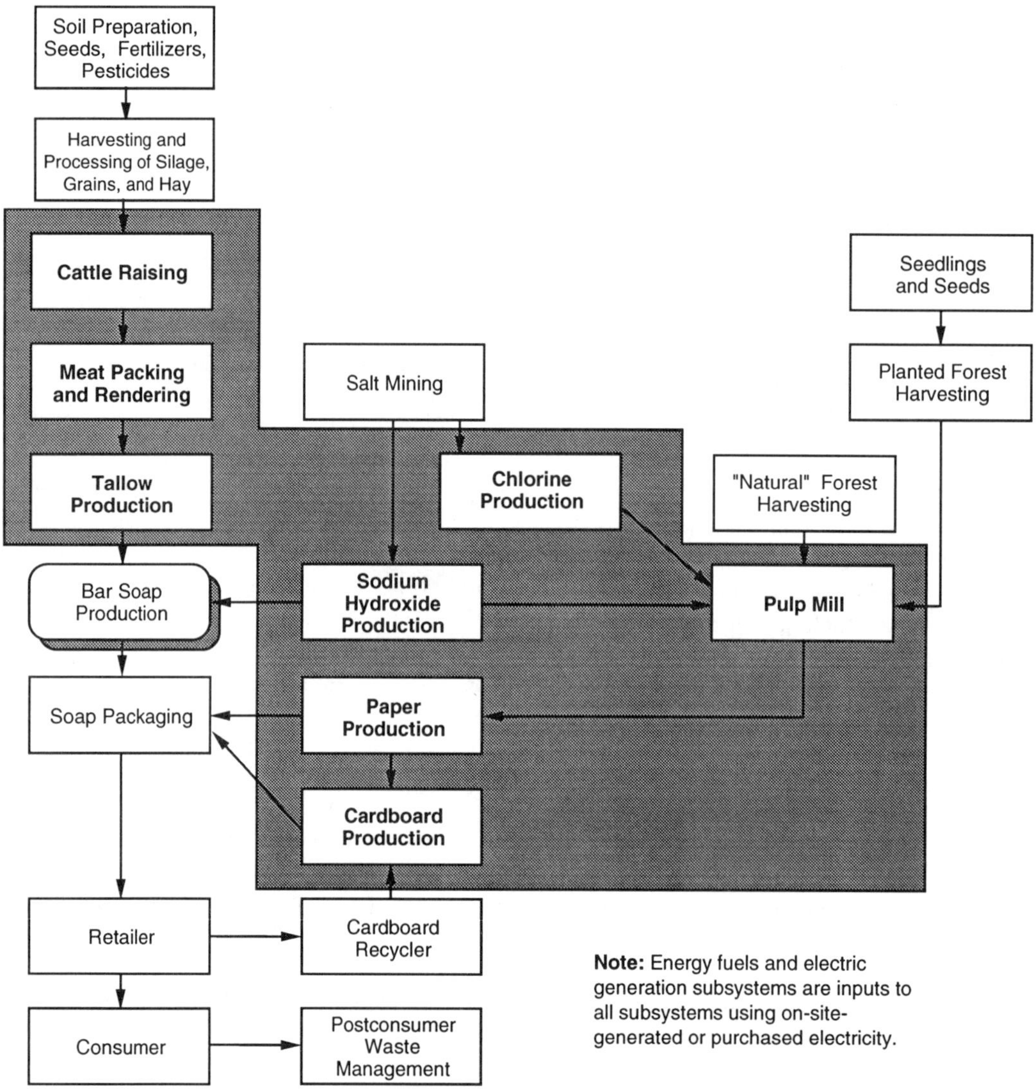

Figure 10
Materials manufacture converts raw materials into a form usable in a finished product

considered to have a continual inner loop of material exiting the system and re-entering the system as an input material. No credits or debits are applied to the processes because the material does not cross subsystem boundaries. For example, when the trim from foam polystyrene extrusion and thermoforming is collected, reground, and used as an input into the subsystem, the amount of raw or intermediate material input from outside the process is reduced. Energy and emissions are calculated based only on the raw or intermediate material input from outside the subsystem.

Material scrap from one manufacturing process also is frequently marketed as a material input in another process. This marketed scrap is commonly referred to as industrial scrap. For example, trim scrap from flexible polyurethane foam often is sold for use as carpet backing. Also, vegetable hulls and peels from food processing operations often are used as raw materials for animal feed. Industrial scrap is viewed as a coproduct from the manufacturing process and carries with it the energy requirements and environmental wastes required to produce that material. Therefore, when the coproduct allocation is applied to the process generating the scrap, the coproduct has the same energy and emissions per pound of output as the operation that uses the scrap in house. Some scrap generated from manufacturing processes is discarded to the municipal solid waste stream along with other wastes from the manufacturing facility. In this case, the scrap is reported as solid waste from the manufacturing process and no allocation is applied.

Product Fabrication Step

The second step in manufacturing is the fabrication of the product. For the manufacture of bar soap, this step is represented by the subsystem that includes the production of fatty acids from tallow, vacuum distillation, manufacture of neat soap and, finally, the cutting and drying of the bar soap.

Subsystem Boundaries

The product fabrication step usually has a more narrow focus than the materials manufacture step and may sometimes include only one manufacturing operation. If more than one fabrication operation is necessary, data may be gathered individually or collectively depending on availability. A product fabrication subsystem will include any energy, material, and water input and the environmental releases. Transport of the product to the point of filling, packaging, and/or distribution is also included. Transportation may occur within the facility or between facilities. The boundaries of the product fabrication step of the life-cycle inventory for bar soap are highlighted in Figure 11. A number of individual processes

Major Concepts—Product Fabrication

- Product fabrication converts intermediate materials into products ready for their intended use by consumers.
- Facilities for which data are reported on a plant-wide basis will require allocation of the inputs and outputs to the product of interest.

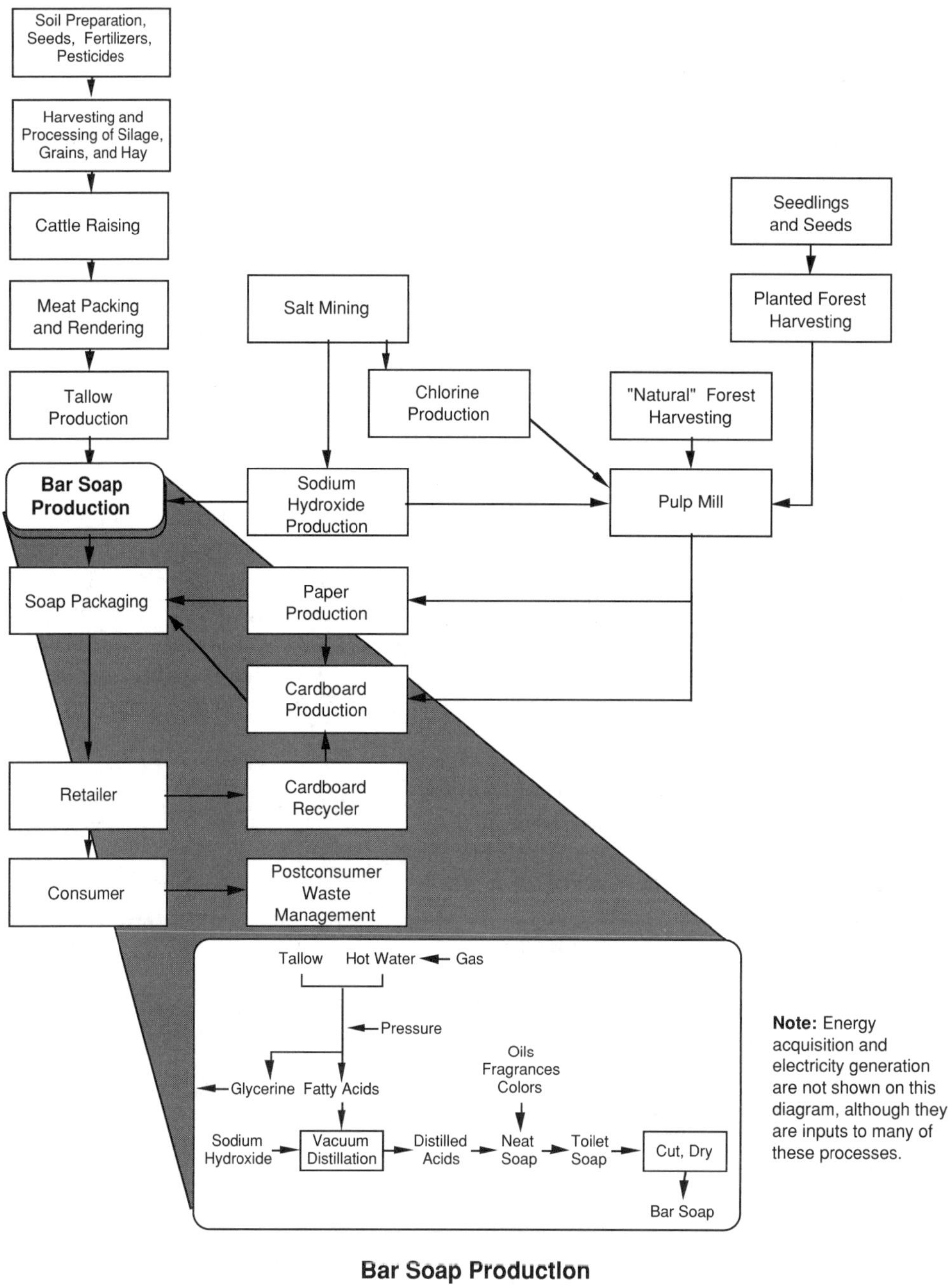

Figure 11
Product fabrication converts intermediate materials into a finished product

are involved, as illustrated by the expanded view in Figure 11. Depending on data availability, the operations can be viewed separately or as comprising a single subsystem. For example, a soap manufacturing plant may account for energy consumption and emissions with plant-wide data rather than process-specific data. In this situation, it would be best to analyze the overall plant operations and then allocate the inputs and outputs. Another soap manufacturing plant may collect and report energy consumption and emissions on each separate process. In this case, each process should be analyzed for its contribution to the entire system or subsystem under examination.

Specific Assumptions and Conventions

The following specific assumptions and conventions generally apply to the product fabrication step:

Coproduct Allocation. Manufacturing processes often produce more than one product. In most cases, only one of those products is evaluated at a time. Thus, some allocation of material and energy inputs and waste and emission outputs is necessary for all marketed output materials. Chapter Four presents several methods for allocating requirements and emissions to various coproducts.

Industrial Scrap. As in the materials manufacture step, scrap material is generated by many product fabrication processes, and the same principles of resource and emissions allocations apply. This marketed scrap is commonly referred to as *industrial scrap*. For example, trim scrap from the production of computers is sold for use in building products.

Filling/Packaging/Distribution Step

The third and final step in the manufacturing stage is filling, packaging, and distribution. This step includes all manufacturing processes and transportation required between product fabrication and delivery of the product to the consumer. Thus, in an inventory that examines bar soap, this step would include all operations required to package the soap in paper wrappers, place the packaged soap into corrugated boxes, and transport the boxes to the retailer and then to the consumer.

Subsystem Boundaries

The subsystem boundaries of the filling, packaging, and distribution step begin with the filling and packaging operations once

Major Concepts—Filling/Packaging/Distribution

- Filling and packaging products ensure that the products remain intact until they are ready for use, whereas distribution transfers the products from the manufacturer to the consumer.
- In addition to primary packaging, some products require secondary and tertiary packaging, all of which should be accounted for in a life-cycle inventory.
- Any special circumstances in transportation, such as refrigeration used to keep a product fresh, should be considered in the inventory.

the materials have reached the facilities where these operations occur. This step also includes distribution to the consumer. Some life-cycle inventories evaluate a package rather than the substance that is put into the package. Therefore, the analysis may not include the contents of the packages. The boundaries of this step of the life-cycle inventory for bar soap are shaded in Figure 12. In this illustration the consumer product, soap, is the focus of the inventory. For each activity or subsystem in this step, the materials and energy inputs are required. Atmospheric and water emissions and the solid wastes resulting from each operation also should be reported. Energy and environmental wastes resulting from transport of the consumer product from the manufacturer and distributor to the retail outlet also are included in this step.

Specific Assumptions and Conventions

The assumptions and conventions discussed below are commonly used for this step of the life-cycle inventory.

Filling. Often the purpose of a life-cycle inventory is to quantify relative differences between the products, processes, or materials being compared. Thus, in filling, certain identical factors between systems often can be ignored because they affect each system in the same way. Two examples of identical factors could be the actual contents of the packages and the filling operation. An example of identical effects from the actual contents of packages would be found in an analysis comparing packages for delivery of 1,000 vitamin tablets. The energy requirements and emissions of the vitamin tablets themselves will always be the same, whereas the energy requirements and emissions of the various package systems being compared may not be the same. Thus the analyst could ignore the tablets and concentrate on the packaging. The energy and emissions associated with filling a product bottle or package can be ignored when the products being compared use the same filling procedures and equipment. For example, a comparison between aluminum soft drink cans and steel soft drink cans will probably have identical filling requirements. However, a comparison between 2-liter plastic soft drink bottles and 12-ounce aluminum cans will have different filling requirements that should be investigated separately in the analysis.

Packaging. Once the bottle or primary package is filled, secondary packaging is applied to ensure the integrity of the product during shipment. The amounts and types of secondary packaging vary with the type of product being shipped. Two-liter plastic soft drink bottles and aluminum cans require different secondary packaging. Bars of soap and liquid soap also require different secondary packaging. When different, the specific amounts and types of secondary packaging should be quantified.

Distribution. Distribution involves transporting the packaged product to warehouses, retail establishments, and consumers. An average distance for product transport must be developed. The normal mode or modes of transportation, e.g., truck, rail, or barge, must be established. Special care should be taken to include the analysis of any unusual situations. For example, a study comparing the delivery of frozen concentrated juice to the delivery of ready-made juice would include the energy and emissions associated with the refrigeration of the frozen product during delivery.

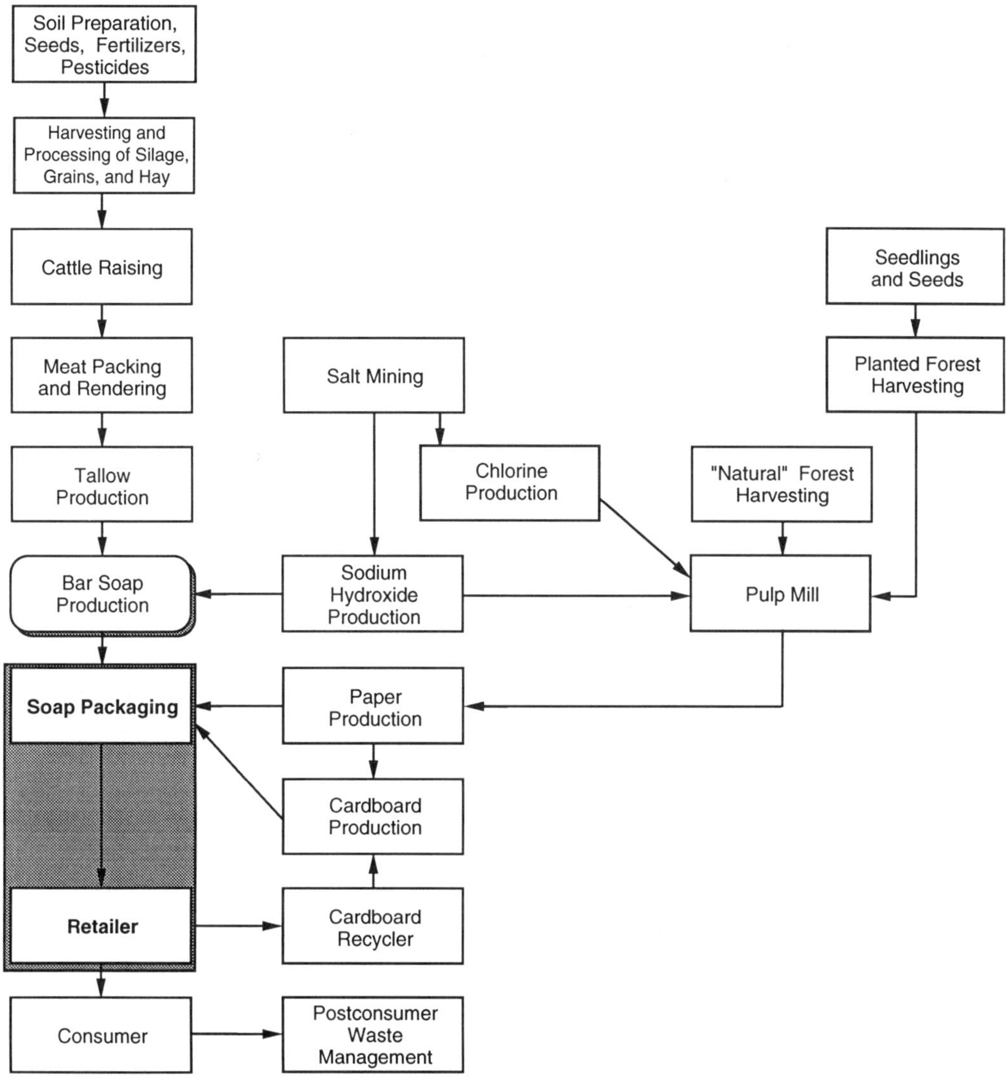

Note: Energy fuels and electric generation subsystems are inputs to all subsystems using on-site-generated or purchased electricity.

Figure 12
Filling, packaging, and distribution is the final step in manufacturing before a product reaches the consumer

USE/REUSE/MAINTENANCE STAGE

The third stage of a product's life cycle is the use/reuse/maintenance (U/R/M) stage. This stage consists of a discrete set of activities that begins after distribution of finished products or materials to the consumer and ends when these products or materials are either recycled or discarded into a waste management system.

Subsystem Boundaries

The U/R/M stage begins when a finished product arrives in the possession of a consumer, including individual consumers, commercial businesses, and institutions. This stage of a life-cycle inventory for bar soap is illustrated by the shading in Figure 13. Transport of the product to the consumer is considered part of the filling/packaging/distribution step, whether it arrives by mail, by purchase from a service outlet, or by transport from a retail store by the consumer, and therefore is not part of the U/R/M stage. The packaging that accompanies the product to the consumer is part of the U/R/M stage, but the shipping boxes used to transport a load of products to the retail store are part of the previous filling/packaging/distribution step. The U/R/M stage ends when the consumer is done using the material or product and delivers it to a collection system for recycling or waste management.

Major Concepts—Use/Reuse/Maintenance

- This stage includes all of the activities undertaken by the user of the product or service as well as any maintenance that may be performed by the user or obtained elsewhere.
- Household operations, such as refrigeration, are rarely associated with a single product. Either the allocation of the capital and operating energy and environmental releases to a particular item are too small to affect the results or they can be proportionately included.

Specific Assumptions and Conventions

Household operations performed during the U/R/M stage are rarely allocated to an individual product. For example, a household rarely operates a refrigerator to cool only one product; household refrigerators usually contain a variety of products. Another example is the variety of clothes that are usually washed or dried in a single load. The historical option for handling the energy requirements and emissions from refrigeration or other similar multiple-product actions has been to omit them from a life-cycle inventory. Sensitivity analysis has shown that once the energy requirements and environmental emissions for these types of multiple-product actions are spread over all products, the values per product are minuscule and do not significantly change the results. A second option for handling multiple-product actions is to proportionally allocate the energy requirements and environmental emissions based on the weight percent, heat capacity, or other justifiable property of each individual product. For example, on an annual basis, "x" loads of laundry are washed, of which

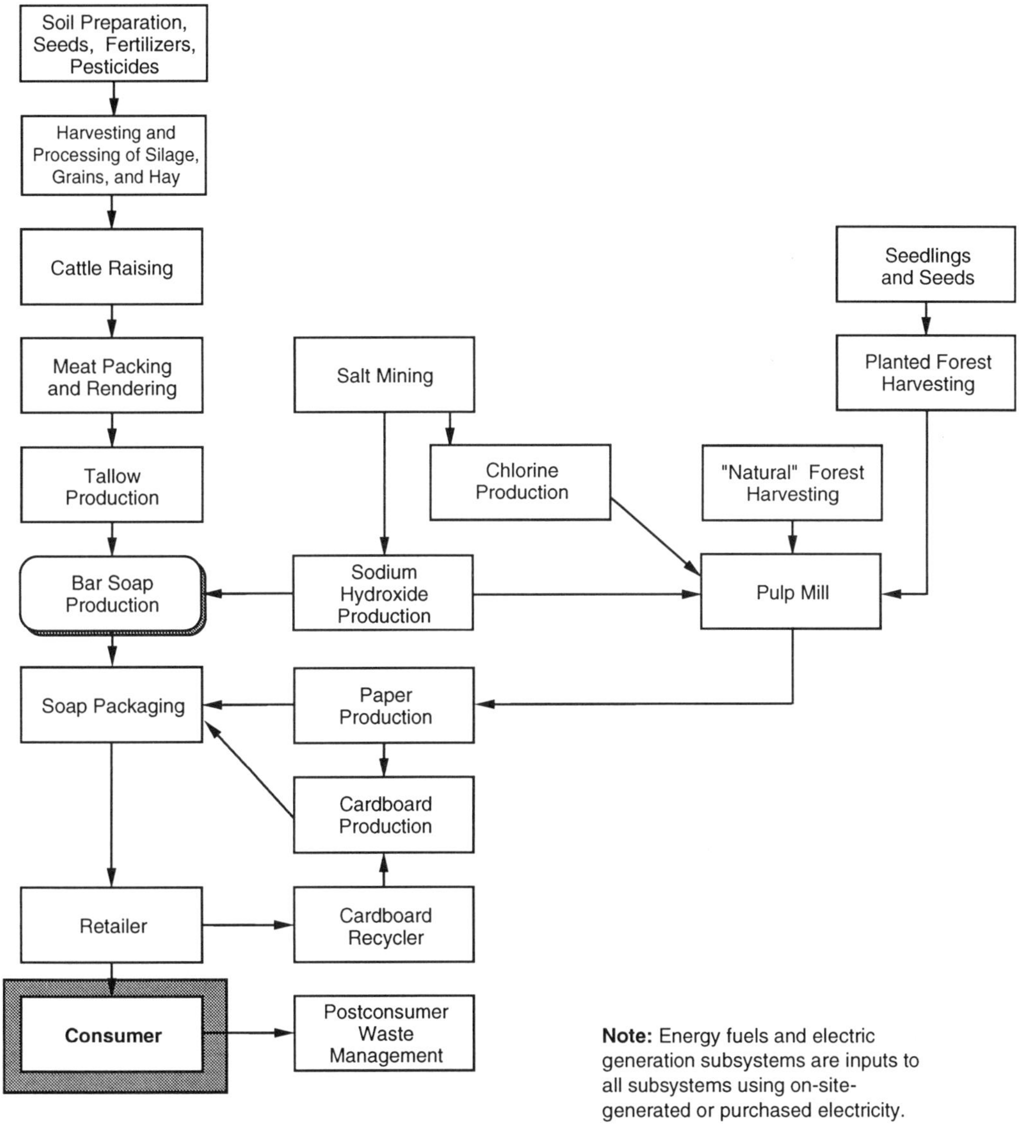

Figure 13
Consumer use/reuse/maintenance is the third stage in a product's life cycle

"y" percent is cloth diapers. The washer-operating emissions allocated to diapers are "z". Over the lifetime of the washer, manufacturing and other emissions are an insignificant percentage of the diaper system life cycle.

The decision to use proportional allocation for consumer products needs to be carefully considered because it can have a significant effect on results. The decision generally relates to the basic purpose of the inventory. In many cases, the purpose of an inventory is to determine the incremental effects of substituting one product for another. For example, if two different bar soap formulations are being compared by a manufacturer, the trip by the consumer in a car to a store to purchase the soap as well as other household goods at the same time would not be included because the effects are identical for either bar soap formulation. If one formulation were to replace the other in the marketplace, no incremental changes in resource depletion or environmental emissions would result.

If, however, the purpose of the inventory were to learn in an absolute sense where resources are used and emissions are generated, the car trip would be included because car trips contribute to global resource depletion and environmental emissions. It is necessary to determine the purpose of the trip because, if the trip is only to purchase one product, the entire trip must be charged to that product. If the trip has multiple uses, proportional allocation would be used.

If the purpose of the inventory were to discover incremental changes by comparing two or more systems, identical consumer actions would be excluded. In absolute studies that include all impacts, consumer activities must be inventoried unless found to be insignificant.

Reuse of a product or package is also included in this stage. A product or package may be reused once or many times. Energy or emissions from activities, such as cleaning, needed to ready the item for reuse should be included in the inventory. If they occur at the point of use, they should be kept in the data for this stage. Sometimes they occur twice, as when a recycled glass container is cleaned by both the user before recycle and by the manufacturer. Some examples of reuse would be refillable glass soft drink bottles and reusable shipping crates. Cleaning or refurbishment energy and emissions should be included with the manufacturing stage data if these operations are not performed by the user.

RECYCLE/WASTE MANAGEMENT STAGE

The fourth and final life-cycle stage is recycle/waste management. Normally, after a product and its packaging have been used by a consumer and the product has fulfilled its intended purpose, it is either recycled, composted, or discarded as waste. Recycling begins when a discarded product or package is delivered to a collection system for recycling. Composting is the controlled, biological decomposition of organic materials into a relatively stable, humus-like material. The waste from the U/R/M life-cycle stage is commonly referred to as postconsumer waste. Waste management refers to the fate of both industrial and postconsumer solid waste that is discarded and picked up, and includes incineration and landfilling. This stage also includes postconsumer wastewater treatment.

Subsystem Boundaries

The subsystem boundaries for recycle/waste management encompass recycling, composting, and the two waste management options of incineration or landfilling for solid waste and wastewater treatment. A variety of steps or subsystems may be included in this stage. A subsystem in this life-cycle stage will include any energy, material, and water inputs and the environmental releases. Transportation from one subsystem to a subsequent one would be included.

The recycle/waste management stage of a life-cycle inventory for bar soap is illustrated by shading in Figure 14. This stage begins with the final disposition of the product and its related packaging. The consumption or use of the product may or may not alter the product itself. The use of soap, for example, reduces the size of the product and thus alters the primary product. The use of disposable diapers also alters the product. Discarded diapers contain added weight and material from human wastes and moisture. Some products are not altered by use. For example, the configuration of a glass beverage container is not changed when the container has fulfilled its purpose of holding a liquid.

The collection and transportation of discarded materials for the various recycling/waste management options should be included in the life-cycle inventory of a product, although typically these steps are minor components. The alternatives most often used in the disposition of discarded products are, in order of EPA preference, reuse, recycling, composting, and incineration/landfilling. The life-cycle inventory should include data for the processing of materials in the recycling and composting processes, i.e., both the energy requirements for these processes and the wastes emitted from them. Incineration converts organic products to carbon dioxide, water, and residuals, but depending on the nature of the product and incinerator control devices, incineration may release atmospheric emissions and/or leave solid and liquid wastes in the forms of ash or scrubber blowdown. The landfilling option ends with the burial of products and requires the quantification of the solid waste buried. Atmospheric and waterborne emissions are also associated with landfilling. Allocation options for these are discussed below.

Major Concepts—Recycle/Waste Management

- Recycle/waste management is the last stage in a product's life cycle.
- In open-loop recycling, products are recycled into new products that are eventually disposed of.
- In closed-loop recycling, products are recycled again and again into the same product.
- Formulas can be used to determine the credits that should be assigned to recycled products analyzed in a life-cycle inventory.

Specific Assumptions and Conventions

This section discusses the basic assumptions and common conventions generally used when performing a life-cycle inventory for the recycle/waste management stage.

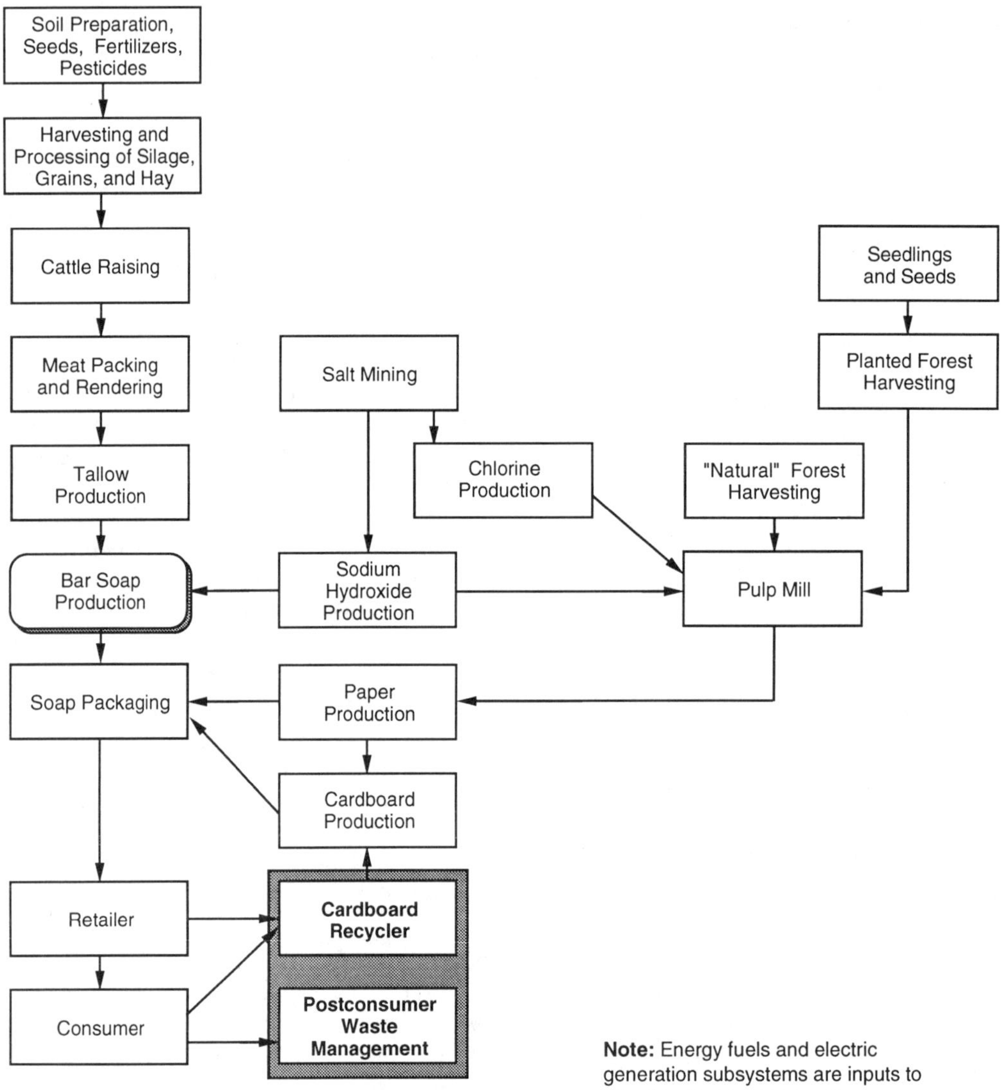

Note: Energy fuels and electric generation subsystems are inputs to all subsystems using on-site-generated or purchased electricity.

Figure 14
Recycle/waste management is the last stage in a product's life cycle

Recycling

Recycling decreases the amount of solid waste entering landfills and reduces the production requirements of virgin or raw materials. Therefore, life-cycle inventory techniques adjust all resource requirements and emissions for products that are recycled or contain recycled content. Two recycling systems could be considered in the life-cycle inventory. These are the closed-loop and open-loop systems.

Closed-Loop Recycling. Closed-loop recycling occurs when a product is recycled into a product that can be recycled over and over again, theoretically endlessly (part a of Figure 15). Aluminum cans are a good example of closed-loop recycling, because they are recycled over and over again into aluminum cans.

Consider an example where virgin materials produced in operation 1 of Figure 15a are augmented with a portion of recycled materials to yield a total mass flow, m, into the production and use stages. If a fraction, F, is recycled through recycling process 4, then $(1-F)m$ is collected from postconsumer use for disposal. A fraction, f, of the material collected for recycling may be rejected, either due to technical reasons (i.e., contamination), or for economic reasons (i.e., low demand creating low recycled material prices), and sent to disposal. Thus, the total disposal mass amounts to $(1-F(1-f))m$. Subtracting the recycling rejects from the recycling input leaves the amount $F(1-f)m$, which is recycled back to the input to augment the new raw material.

If no recycling takes place, then both the raw materials input mass and the waste disposal mass are equal to the value m. It can be seen that as the recycled fraction, F, increases toward unity, the need for virgin raw materials and the solid waste generation rate asymptotically approach the recycling reject rate of fm at 100% recycling. Ultimately, raw material resource use and solid waste disposal reach zero as the efficiency of the recycling process approaches 100%. Thus, for closed-loop recycling, allocation of inputs and outputs is straightforward once the recycling rate and the recycling process reject rate are known.

Open-Loop Recycling. In the basic open-loop recycling system, a product made from virgin material is recycled into another product that is not recycled, but disposed of, possibly after a long-term diversion. An example of open-loop recycling would be a plastic milk bottle being recycled into plastic lumber or flower pots, which currently are not recycled. In open-loop recycling, energy and environmental emissions related to the production, recycling, and final disposition of the plastic resin itself are divided between the two products by one of several possible methods. In this way, each of the two products made from the same resin shares in the energy, landfill, and air and

Examples of Processes in Postconsumer Recycling

Plastic milk jugs

- Grinding
- Washing
- Remelting

Paper products

- Repulping and deinking/bleaching

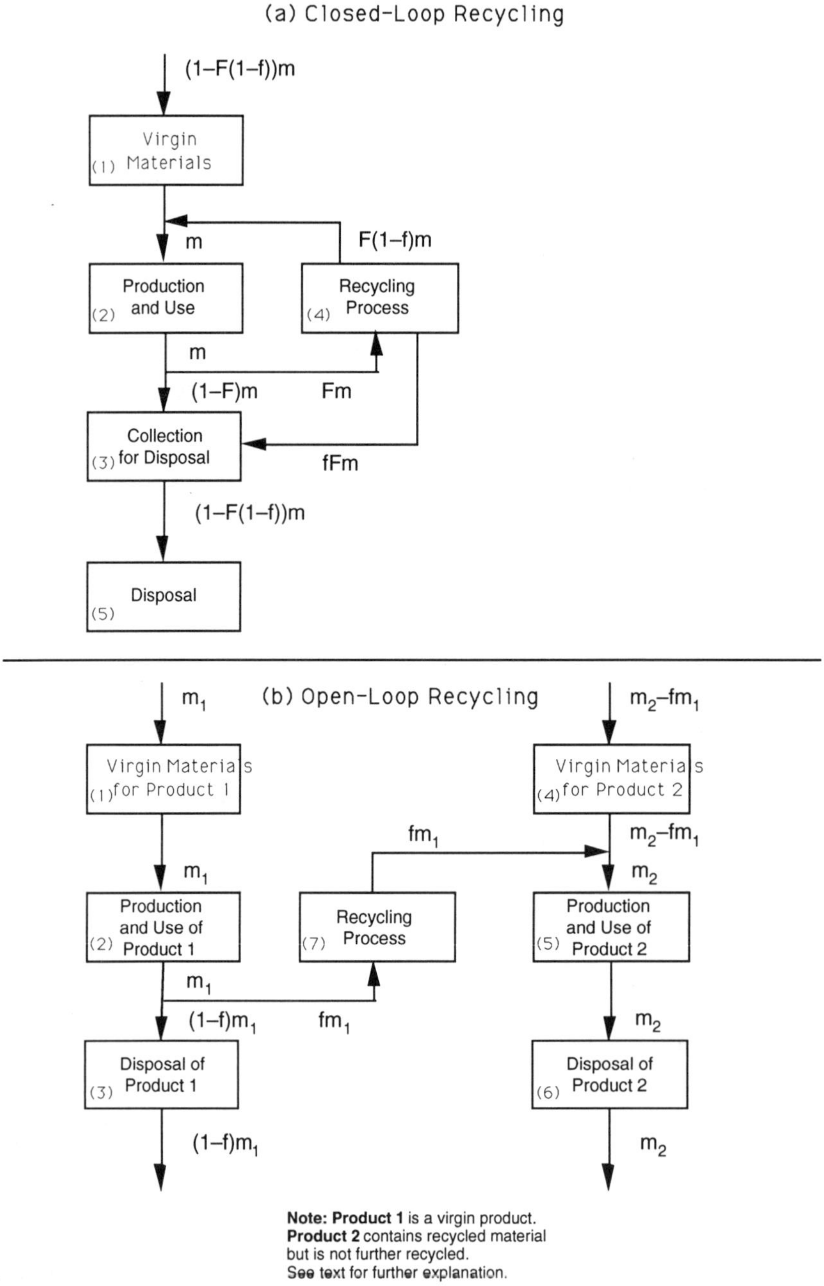

Figure 15
Recycling flow diagrams

water emissions savings achieved through recycling. This basic open-loop system is illustrated in part (b) of Figure 15.

Analysis of an open-loop system is only slightly more complicated from a mathematical perspective than that for a closed-loop system. In the simplest open-loop system, Product 1's production sequence produces an output from operation (2) that is diverted from waste disposal by some fraction f. Thus, the mass flow to the recycling operation is fm_1. Again, although not shown, a fraction can be diverted from collection to disposal to account for poor quality or economics just as in the closed-loop case. Product 2 has its virgin raw materials input reduced from m_2 to $m_2–fm_2$. After use, Product 2 is disposed of, with no further recycling, at mass rate m_2. As the recycling rate of Product 1 approaches 100%, the solid waste generation tends toward zero and the raw materials input to Product 2 approaches $m_2–m_1$.

Correct allocation of the inputs and outputs requires analyzing the systems for both Product 1 and Product 2 together. This is the preferred option for practitioners to pursue. However, in situations where this option is not feasible, there are several possible methods for allocating all energy and water requirements and environmental releases attributable to the products (except those for fabrication, packaging, and distribution). Three possible methods for allocating impacts between the two products are discussed below.

Because all of the following three allocation methods for open-loop recycling are arbitrary, they are listed in decreasing order of complexity to analyze: (1) the system impacts can be allocated between the two products based on the percentage of the two products produced; (2) the disposal credits can be allocated to the product being recycled; and (3) the impacts added to the system because of recycling can be equally divided. No matter which allocation method is used, it is important for the analyst to state clearly the method being used, to explain why it was chosen, and to use it consistently. For each of these allocation methods, the *net* inputs and outputs ($N_{I/O}$) can be described generically by the following formula:

$$N_{I/O} = D + E - B(a) - A(b) + C(a) + F$$

where:

A = all inputs and outputs associated with production of the virgin material from raw material to primary material production for Product 2 (the product that uses the recycled material).

B = all inputs and outputs associated with disposal of Product 1 (the product that is recycled), including transportation, solid waste, incineration emissions, and all other system impacts associated with waste disposal.

C = all inputs and outputs associated with recycling of Product 1, including transportation (to a drop-off site, a materials recovery system, or a processor); energy to reprocess back into a primary material (to clean, shred, or repulp); water; and wastes associated with this reprocessing.

D = all inputs and outputs associated with a no-recycling system for Product 1.

E = all inputs and outputs associated with a virgin system for Product 2.

F = any converting inputs and outputs incurred as a result of Product 2 using recycled materials instead of virgin materials.

a = the recycling rate of Product 1.

b = the recycled content level of Product 2.

The inputs and outputs represented by D, E, B, A, C, and F in this equation are specific quantities associated with a given mass of either Product 1 or Product 2. For example, if 200 pounds of Product 1 and 100 pounds of Product 2 are the result of the system, then D, B, and C quantities are for 200 pounds of Product 1 and E, C, and F quantities are for 100 pounds of Product 2. Inputs and outputs represented by $N_{I/O}$ are for the entire system or 300 pounds of products.

First Allocation Method. In the first allocation method, the system loadings can be allocated between the two products based on the number of units of each product needed to produce a combined total of 100 units of both products. This allocation method requires that both product systems be analyzed to determine the percent of Product 1 that is recycled and the recycled content level (from Product 1) in Product 2. The net inputs and outputs can be calculated by the following formulas for each product:

$$N_{I/O}\ (\text{Product 1}) = D - P_1\ [B(a) + A(b) - C(a) - F]$$

$$N_{I/O}\ (\text{Product 2}) = E - P_2\ [B(a) + A(b) - C(a) - F]$$

where:

P_1 = the percent of Product 1 out of 100 total units for both products.

P_2 = the percent of Product 2 out of 100 total units for both products.

The limitations of this method are that Product 2 may not be fully credited for recycle content. For example, in a system where Product 1 is being recycled at 50% into Product 2, which has 100% postconsumer recycle content from Product 1, Product 1 will be credited with higher "savings" from recycling than will Product 2 because Product 1 makes up a larger percentage of the system. Although Product 2 is made from 100% recycled material, it is penalized because Product 1 is being recycled at only 50%.

Second Allocation Method. In the second allocation method, the disposal credits are allocated to the product that gets recycled (Product 1). Product 2 treats Product 1 as a raw material that needs to be processed; thus, the recycling inputs and outputs are all allocated to Product 1. By treating Product 1 as a raw material, Product 2 has avoided virgin material production. The net inputs and outputs for this allocation method can be calculated by the following formulas for each product:

$$N_{I/O}\ (\text{Product 1}) = D - B(a)$$

$$N_{I/O}\ (\text{Product 2}) = E - A(b) + C(a) + F$$

This method has the following limitations: Product 2 is penalized with all of the solid waste from the system, when actually the production of Product 2 using recycled material has decreased the system's total solid waste by the quantity of material being recycled. Product 1 receives no credit for virgin materials savings even though the availability of Product 1 for recycling reduces virgin material manufacture for making Product 2. However, this allocation method is very useful when primary materials and their productions are different or for composted materials.

Third Allocation Method. In the third allocation method, the loadings added to the system because of recycling are divided equally between the two products. The loadings associated with recycling are reduced disposal of Product 1, reduced virgin material product for Product 2, inputs and outputs associated with recycling, and any converting net inputs and outputs incurred as a result of using recycled materials over virgin materials in Product 2. The net inputs and outputs for this 50/50 allocation method can be calculated by the following formulas for each product:

$$N_{I/O} \text{ (Product 1)} = D - 1/2[B(a) + A(b) - C(a) - F]$$

$$N_{I/O} \text{ (Product 2)} = E - 1/2[B(a) + A(b) - C(a) - F]$$

This allocation method has the following advantages: It eliminates the possibility of double-allocating recycling impacts, minimizes arguments over which product should receive the recycling impacts, and enables independent evaluation of a product. The limitation of the 50/50 allocation method is that it does not give credit to the party (e.g., manufacturer, recycler, municipality) that made the effort to implement the recycling change.

For all three methods typically, the impacts of virgin material on both products must be assumed to be the same, because a manufacturer may have information on only the virgin material for the product made in house. The assumption of virgin material similarity between the two products may be true for some cases of recycling, such as recycling milk bottles into detergent bottles, but may not be true for other recycling situations, such as recycling of some paper products.

Composting

Composting is an alternative to disposal by which organic materials are removed from the solid waste stream destined for the landfill. Composting is the controlled, biological decomposition of organic materials into a relatively stable humus-like product, which can be handled, stored, and/or applied to the land without adversely affecting the environment (BioCycle, 1991). The compost produced from the process has only modest nutrient value, but is considered an excellent soil amendment. Composted materials can replace peatmoss, reduce the use of topsoil, agrichemicals, water, and sometimes fertilizer, be used as animal bedding, replace landfill cover material, etc. Thus, composting can be viewed as a form of open-loop recycling. The process decreases the volume of discarded material occupying landfills; thus, when composting is considered in a life-cycle inventory, the solid waste for a given material is reduced. Composted products may contain noncompostable materials; thus, noncompostable solid waste resulting from the composting of a product is attributed to that product in a life-cycle inventory. For example, foil found in some paperboard containers is not compostable. It is, therefore, part of the solid waste screened out of the composted product and would be attributed to the solid waste component of the paperboard containers.

Emissions from Solid Waste Management Options

The incineration and/or landfilling of mixed municipal solid waste (MSW) creates problems in determining the atmospheric or waterborne emissions from such processes for particular products. For example, how does one quantify the contribution a milk bottle would make to the leachate collected from a landfill compared to the contribution of a newspaper? Or the amount of incinerator emissions resulting from the combustion of a plastic film compared to the combustion of a paperboard box? Little research on specific products or even specific materials has been done. The primary historical option has been to omit reporting the emissions from solid waste management processes, due to the difficulty in accurately reporting the emissions from the combustion of MSW or from leachate and air emissions from mixed solid waste in a landfill.

Whenever a product disposition option specific to a particular product is used, emissions from the disposition process are quantifiable for that product and can be reported. However, currently no accurate method exists to allocate incinerator or landfill emissions to a particular product once it has been combined with other mixed solid waste. For example, composting programs for yard waste only and diapers only are able to collect and report emissions for the composting of those particular materials. Recycling processes also are nearly always specific to a particular material. Aluminum, paper, plastics, and glass require separate processes to convert discarded products into usable materials.

Nevertheless, an attempt should be made to account for MSW emissions in a life-cycle inventory whenever possible. Several potential approaches exist for estimating these emissions, but additional research is needed to establish the accuracy and utility of these methods. For almost any of these approaches, it is important to know the percent of a particular waste that is recycled, incinerated, or landfilled. These waste disposal percentages are available on a national basis for typical MSW, but they may not be relevant for specific products. For example, some products have a high recycling rate although the generic materials may not, and other products are not recycled at all although the general material type may be. Also, emissions data from a typical MSW incinerator or landfill are available, but the materials in the mixed waste causing these emissions are not specified. One option for using both the waste disposal percentages and the typical emissions from MSW incinerators or landfills is to proportionally allocate the emissions based on a broad input material composition. The range of selected landfill leachate descriptors has been reported in Chian et al. (1986). Landfill emissions associated with a few specific types of materials also are available, e.g., dry cell batteries (Jones et al., 1977) and plastics (Wilson et al., 1982).

Lacking even this information, emissions from a "typical" facility could be allocated based on the weight percentage of MSW that the product comprises. Similarly, incinerator emissions associated with a few specific materials are available, e.g., polyvinylchloride (PVC) (Carroll, 1988). The major groups of incinerator air emissions listed by SETAC (Fava et al., 1991) include CO_2 and H_2O, criteria pollutants (NO_x, SO_x, particulates, VOCs, lead, and CO), products of incomplete combustion (PICs) and particulate organic chemicals (POCs), heavy metals, dioxans and furans, and waste heat.

A second option is to make idealized estimates of emissions associated with waste management methods by simulation modeling. Although the modeling methods have not been validated by trial burns for very many materials, incinerator emissions may be estimated by equilibrium or stoichiometric methods.

A third option for estimating emissions associated with the incineration or landfilling of a particular product is to analyze the leachate or air emissions composition coming from a mixed waste and assign emissions to the product based on degradation product fingerprinting. This option would be highly resource intensive.

Energy from Incineration

Many products release energy when burned in an incinerator. This energy can and has been credited to a product, reducing the net reported energy requirements for the life cycle of the product. The energy credit traditionally is determined using the higher heating value (HHV) of the material, taking into account the moisture content of the waste as disposed. The value of the recovered energy for products manufactured from fossil fuels, e.g., plastics, will not equal the energy of the raw material resource attributed to the product, because the change in chemical structure changes the recoverable amount of energy.

In principle, it should be possible to employ a second option if the actual delivered energy is calculated by adjusting for moisture as well as system thermal losses, e.g., heat transfer to walls and thermodynamic conversion losses.

REFERENCES

BioCycle, 1991. The Art and Science of Composting. Edited by the staff of BioCycle, Journal of Waste Recycling. J.G. Press, Inc., Emmaus, PA.

Boustead, I. and G. F. Hancock, 1979. Handbook of Industrial Energy Analysis. Chichester: Ellis Horwood and New York: John Wiley, ISBN 0-470-26492-6, Chapter 3, "Real Industrial Systems," p. 76.

Boustead, I., undated. "The Relevance of Re-Use and Recycling Activities for the LCA Profile of Products," 10 p.

Brown, H.L., B.B. Hamel, B.A. Hedman, et al., 1985. Energy Analysis of 108 Industrial Processes, Drexel University, Philadelphia, PA. Prepared for U.S. Department of Energy. Fairmont Press, 314 p.

Canadian Electric Utilities and National Energy Board, 1992. Personal Communication between Raffaele DiGirolamo, Energy, Mines, and Resources Canada, and Illa Amerson, Battelle.

Carroll, W.F., Jr., 1988. "PVC and Incineration." J. Vinyl Technol. 10(2):90-94.

Chian, E.S.K., S.B. Ghosh, B. Kahn, M. Giabbi, and F.G. Pohland, 1986. Codisposal of Low Level Radioactive Waste and Sanitary Waste, SCEGIT Report No. 86-01.

DOE (U.S. Department of Energy), 1992, "Monthly Power Plant Report," Energy Information Administration, EIA-759.

Fava, J.A., R. Denison, B. Jones, M.A. Curran, B. Vigon, S. Selke, and J. Barnum (Editors), 1991. A Technical Framework for Life-Cycle Assessments. Society of Environmental Toxicology and Chemistry and SETAC Foundation for Environmental Education, Inc., Washington, D.C.

Fava, J.A., R. Denison, T. Mohin, and R. Parrish, 1992. "Life-Cycle Assessment Peer Review Framework." Society of Environmental Toxicology and Chemistry, Life-Cycle Assessment Advisory Group, 4 p.

Jones, C.J., P.J., McGugan, and P.F. Lawrence, 1977. "An Investigation of the Degradation of Some Dry Cell Batteries Under Domestic Waste Landfill Conditions." J. Hazard. Mater., 2:259-289.

Jorgensen, M.S. and B. Pedersen, 1990. "Quality Concepts for Processed Organic Food." Letter Attachment to Timothy Mohin, U.S. Environmental Protection Agency from Bo Pedersen, Interdisciplinary Centre, Technical University of Denmark, Lyngby, Denmark, October 17, 1990.

Lundholm, M.P. and G. Sundstrom, 1985. "Tetra Brik Aseptic Environmental Profile," 174 p.

Meadows, D. H. et al., 1972. The Limits to Growth; a Report for the Club of Rome's Project on the Predicament of Mankind. Universe Books, New York. 205 p.

Raiffa, H., 1968. Decision Analysis—Introductory Lectures on Choices under Uncertainty. Addison-Wesley, Reading, MA.

Sauer, B.J., R.G. Hunt, and M.A. Franklin, 1990. "Background Document on Clean Products—Research and Implementation." U.S. Environmental Protection Agency, Risk Reduction Engineering Laboratory, Cincinnati, OH. EPA/600/2-90/048. 75 p.

Tillman, A.M., H. Baumann, E. Eriksson, and T. Rydberg, 1991. "Life-cycle Analyses of Selected Packaging Materials: Quantification of Environmental Loadings." Report from Chalmers Industriteknik to the Swedish Commission on Packaging, 206 p.

Werner, A.F., 1991. "Product Lifecycle Assessment: A Strategic Approach." Proceedings of the Global Pollution Prevention '91 Conference, Washington D.C.

Wilson, D.C., P.J. Young, B.C. Audson, and G. Baldwin, 1982. "Leaching Cadmium from Pigmented Plastics in a Landfill Site." Environ. Sci. Tech. 16(9):560.

GLOSSARY

Accidental emission: An unintended environmental release. Examples: crude oil spills resulting from tanker accidents, venting of reactors due to mechanical failure or human error.

Atmospheric emissions: Residual discharges of emissions to the air (usually expressed in pounds or kilograms per unit output) following emission control devices. Includes point sources such as stacks and vents as well as area sources such as storage piles.

Biomass: The weight of living material, occasionally used as an energy source.

Brines (oilfield): Wastewater produced along with crude oil and natural gas from oilfield operations.

Btu (British thermal unit): The quantity of heat energy required to raise the temperature of 1 pound of water (air-free) from 60° to 61° Fahrenheit at a constant pressure of 1 standard atmosphere. Experimentally equal to 1,054.5 joules.

By-product: A useful product that is not the primary product being produced. In life-cycle analysis by-products are treated as coproducts.

Closed-loop recycling: A recycling system in which a particular mass of material is remanufactured into the same product (e.g., glass bottle into glass bottle).

Composite data: Data from multiple facilities performing the same operation that have been combined or averaged in some manner.

Composting: A waste management option involving the controlled biological decomposition of organic materials into a relatively stable humus-like product that can be handled, stored, and/or applied to the land without adversely affecting the environment.

Consumer use: The intended end use of a product. The use for which a product was designed.

Coproduct: A marketable by-product from a process. This includes materials that may be traditionally defined as wastes such as industrial scrap that is subsequently used as a raw material in a different manufacturing process.

Coproduct allocation: Adjustment of material inputs, energy requirements, and environmental emissions from a process to allocate those impacts attributable to the output product being considered.

Energy characterization: Classification of energy according to primary fuel source: wood, natural gas, petroleum, coal, nuclear, hydropower.

Energy of material resource: The fuel value of the raw materials used to make a product. The inherent energy in a product made from a raw material used as a fuel supply. Also known as latent energy.

Energy profile: A listing of the energy usage for a system by stage and/or by source. See also energy characterization.

Environmental loadings: The releases of pollutants to the environment.

Environmental release: Emissions or wastes discharged to the air, land, or water. Contaminants that cross a system boundary into the environment.

Equivalent usage ratio: A method of comparing two or more different products on an equivalent basis. For example, a comparison of beverage containers based upon the quantity of beverage delivered to the consumer.

Error analysis: A systematic method for analyzing differences between a measured or estimated quantity and the true value.

Finished product: The product produced by the system being evaluated for consumer purchase.

Fuel-related emissions: See fuel-related wastes.

Fuel-related wastes: Those materials or emissions generated during the combustion of fuels for the production of heat, steam, electricity, or energy to power processes and transportation equipment that are not a component of the useable product or coproducts.

Fuel unit: Weight or volume of fuel such as gallons of fuel oil, pounds of coal, or cubic feet of natural gas.

Fugitive emissions: Emissions from valves or leaks in process equipment or material storage areas that are difficult to measure and do not flow through pollution control devices.

Gigajoules (GJ):	1,000,000,000 or 10^9 joules.
Global warming:	The theory that elevated concentrations of certain atmospheric constituents are causing an increase in the earth's average temperature.
Greenhouse effect:	The theory that certain atmospheric constituents trap heat in the earth's atmosphere leading to global warming.
Greenhouse gas:	An atmospheric constituent, such as carbon dioxide, that is thought to contribute to global warming.
Higher heating value (HHV):	The gross heat of combustion for a material.
Impact analysis:	The assessment of the environmental consequences of energy and natural resource consumption and waste releases associated with an actual or proposed action.
Improvement analysis:	The component of a life-cycle assessment that is concerned with the evaluation of opportunities to effect reductions in environmental releases and resource use.
Incineration credit:	The energy credit given in a life-cycle inventory for the burning of material in a waste-to-energy incinerator.
Industrial scrap:	The waste materials of value produced by a manufacturing process. The material is often reused within the same process. It may also be sold to another operation as a raw material.
Industrial solid waste:	Industrial solid waste includes wastewater treatment sludges, solids from air pollution control devices, trim or scrap materials that are not recycled, fuel combustion residues (such as the ash generated by burning wood or coal), and mineral extraction residues.
Inherent energy:	See Energy of Material Resource.
Intermediate materials:	The materials made from raw materials and from which finished products are made.
Inventory analysis:	See Life-cycle Inventory.
Joule:	SI (metric) unit of energy, equal to 9.48×10^{-4} Btu.
Leachate:	The solution that is produced by the action of percolating water through a permeable solid, as in a landfill.
Life cycle:	The stages of a product, process, or package's life, beginning with raw materials acquisition, continuing through

processing, materials manufacture, product fabrication, and use, and concluding with any of a variety of waste management options.

Life-cycle assessment: A concept and a methodology to evaluate the environmental effects of a product or activity holistically, by analyzing the entire life cycle of a particular product, process, or activity. The life-cycle assessment consists of three complementary components—inventory, impact, and improvement—and an integrative procedure known as scoping.

Life-cycle inventory: The identification and quantification of energy, resource usage, and environmental emissions for a particular product, process, or activity.

Life-cycle stages: The stages of any process, including raw materials and energy acquisition; manufacturing (including materials manufacture, product fabrication, and filling/packaging/distribution steps); use/reuse/maintenance; and recycle/waste management.

MJ value: Megajoule, 1,000,000 joules, equal to 948 Btu. See Joule.

Model (computational): A computational framework, usually a computer spreadsheet or other such tool, that incorporates the stand-alone data and materials flows into the total results for the energy and resource use and environmental releases from the overall system.

Municipal Solid Waste (MSW): MSW includes wastes such as durable goods, nondurable goods, containers and packaging, food wastes, yard wastes, and miscellaneous inorganic wastes from residential, commercial, institutional, and industrial sources. MSW does not include wastes from other sources, such as municipal sludges, combustion ash, and industrial nonhazardous process wastes that might also be disposed of in municipal waste landfills or incinerators.

National electricity grid: The electricity generated by individual generators nationally that are interconnected to form regional grids and also a national grid.

Nonrenewable resource: A resource that cannot be replaced in the environment as fast as it is being consumed.

Open-loop recycling: A recycling system in which a product made from one type of material is recycled into a different type of product. The product receiving recycled material itself may or may not be recycled.

Overburden: The material to be removed or displaced that is overlying the ore or material being mined.

Packaging, primary: The level of packaging that is in contact with the product. For certain beverages, this might be the 12-ounce aluminum can.

Packaging, secondary: The second level of packaging for a product that contains one or more primary packages. For 12-ounce beverage cans, this might be the plastic rings to hold the 6-pack together.

Packaging, tertiary: The third level of packaging for a product that contains one or more secondary packages. For 6-packs of 12-ounce beverages cans, this might be the corrugated trays and stretch wrap over the pallet that are used in transporting the product.

Postconsumer solid waste: A material that has served its intended use and has become a part of the waste stream.

Precombustion energy: Energy required to extract, transport, and process the fuels used for power generation. Includes adjustment for inefficiencies in power generation and for transmission losses. Also known as energy of fuel acquisition.

Predominant industrial practice: A practice generally acknowledged to be widely used as a significant percentage either of companies in the industry or of the total material flow.

Process: The operations that make up subsystems.

Process emissions: Waste materials generated or produced from the raw materials, reactions, processes, or related equipment inherent to the process.

Process energy: The energy required for each subsystem for process requirements. These are quantified in terms of fuel or power units such as gallons of distillate oil, cubic feet of natural gas, or kilowatt-hours (kWh) of electricity.

Process-related wastes: The waste materials generated or produced from the raw materials, reactions, processes, or related equipment inherent to the process.

Random error: Error that does not have any definite or systematic pattern or bias.

Raw materials: The total inputs for a subsystem including all material present in the product and material found in losses due to emissions, scrap and off-spec products, and no-emission losses (such as moisture due to evaporation). Water is not always a raw material input because it is often removed during a drying step.

Recycled content: The amount of recovered material, either pre- or postconsumer, in a finished product that was derived from materials diverted from the waste management system. Usually expressed as a percent by weight.

Regional electricity grid: The mix of fuel sources used to generate electricity for a given region. A regional electricity grid is occasionally used in a life-cycle inventory when an energy-intensive industry is sited in a certain region to take advantage of inexpensive electric power.

Regulated emissions: Those emissions regulated by government to limit amounts or concentrations of waste.

Renewable resource: A resource that can be replaced in the environment faster than it is being consumed.

REPA: Resource and Environmental Profile Analysis. Also commonly called cradle-to-grave analysis or life-cycle analysis.

Representativeness: The state of being a sample that is characteristic of a group or population of operations or processes.

Residual oil: The heavier oils that remain after the distillate fuel oils and lighter hydrocarbons are removed in refinery operations. Included are No. 5 and No. 6 oils.

Residues: Process wastes (such as wood bark), typically but not always solids.

Resource requirements: The amounts of raw materials or natural inputs and energy used in a system.

Risk assessment: An evaluation of potential consequences to humans, wildlife, or the environment caused by a process, product, or activity and including both the likelihood and the effects of an event.

Routine emissions: Those releases that normally occur from a process, as opposed to accidental releases that proceed from abnormal process conditions.

Sensitivity analysis: A systematic evaluation process for describing the effect of variations of inputs to a system on the output.

SI (Système Internationale): Internationally used system of standards for units and dimensions.

Soil amendment: The product of composting of organic materials that is applied to the soil as a conditioning agent.

Solid waste: Solid products or materials disposed of in landfills, incinerated, or composted. Can be expressed in weight or volume terms.

Specific data: Data that are characteristic of a particular subsystem or process.

Stand-alone data: Normalized data consistently defining the system by reporting the same product output from each subsystem (e.g., on the basis of 1,000 pounds of output).

Subprocess: An individual step that is a part of a defined process.

Subsystem: An individual process that is a part of the defined system.

System: A collection of operations that perform a desired function. In a life-cycle inventory, the scope of the system is defined by the boundary conditions.

Systematic error: Error that is not random. Variation caused, for example, by differences in age of equipment, advances in technology, or local conditions.

Systems analysis: A stepwise evaluation of the inputs and outputs of a defined system.

Template: A guide used by analysts for collecting and organizing data. The template describes a material and energy balance for a defined system or subsystem. It includes resource requirements, transportation requirements, and emissions and wastes for that system or subsystem.

Ton-mile: A measure of the movement of 1 ton (2,000 pounds) of freight for the distance of 1 mile. For example, 100 ton-miles is the measure for moving 100 tons of freight 1 mile. It could also represent moving 1 ton of freight 100 miles.

Transmission line loss: The difference between electricity generated and electricity delivered.

Transportation energy: Energy required to transport materials and products throughout the process and to final distribution to the consumer. This is converted from the conventional units of "ton-miles" by each transport mode (e.g., truck, rail, barge, airfreight, pipeline, etc.) using on the average efficiency of each mode.

Waterborne wastes: Discharges to water of regulated pollutants (usually expressed in kilograms per unit output) after existing treatment processes.

APPENDIX

Clean Air Act (CAA)

The Clean Air Act (CAA) is a piece of national legislation designed to identify and control pollutants and sources of emissions that may reduce the quality of the nation's air. The CAA has been amended twice since its inception in an effort to adjust it to our changing perception and knowledge of the environment. The most recent amendments in 1990 identify 189 toxic substances and pollutants as well as sources of emissions that enter the nation's air. The objective of the CAA is to restore and maintain the chemical, physical, and biological integrity of the nation's air.

Under CAA Title III, industrial facilities are subject to new source performance standards for new facilities to be constructed or notification of changes to existing ones after the date the EPA proposes new source performance standards (40 CFR Part 60).

There are 174 categories of sources listed pursuant to CAA Title III. These include steam generators (both fossil fuel and petroleum), incinerators, cement plants, chemical acid plants, petroleum refineries, sewage treatment plants, metal smelting facilities, fertilizer production plants, steel plants, paper mills, glass manufacturing plants, synthetic fiber production facilities, synthetic organic chemical manufacturing, and nonmetallic mineral processing plants.

In CAA subpart A, section 61.01, a list of pollutants deemed hazardous and their applicability has been published pursuant to CAA section 112 (CAA-5).

In CAA Part 61, a list of National Standards has been created to control those emissions deemed hazardous air pollutants. Examples of the emissions are radon from uranium mines, beryllium, mercury, vinyl chloride, radionuclides other than radon, benzene leaks, radionuclide emissions from phosphorus plants, asbestos, and inorganic arsenic emissions from glass manufacturing and primary copper smelting.

A priority list of major source categories is given in CAA section 60.16. Among the largest sources of hazardous pollutants are synthetic organic chemical manufacturing, petroleum refineries, dry cleaning, graphic arts, stationary combustion engines, and industrial surface coating of fabric. In all there are 59 sources; however, several of these sources are no longer considered priority sources, i.e., mineral wool, secondary copper, and ceramic clay manufacturing.

Title I of the CAA Amendments of 1990 extends and revises the existing requirements for attaining and maintaining the National Ambient Air Quality Standards (NAAQS) for the following six criteria pollutants: ozone, carbon monoxide (CO), particulate matter (PM_{10}), lead (Pb), sulfur dioxide (SO_2), and nitrogen oxides (NO_x). Title I also addresses permit requirements and emissions inventories for existing stationary sources. The primary industry segments impacted by Title I include chemicals, pulp and paper, petrochemicals, pharmaceuticals, iron and steel, and most manufacturing industries. Title I also changed the definition of major stationary source relative to volatile organic carbon emissions (VOCs).

Title II of the CAA Amendments of 1990 covers mobile emission sources. The requirements affect tailpipe emission standards for CO_2, hydrocarbons, and CO.

Title III of the CAA Amendments of 1990 established emission standards for 189 air toxics, or hazardous air pollutants (HAPs), such as acrylonitrile or chlorine. The list of HAPs is divided into industry groups, such as polymer and resin production, and then identifies individual source categories, such as cellophane or polystyrene production. Title III also provides for comprehensive regulation of solid waste incinerators and for development of a list of at least 100 HAPs which, if released accidentally, could seriously threaten human health or the environment.

Title IV of the CAA Amendments of 1990 seeks to reduce annual emissions of SO_2 and NO_x, because they are principal components in the formation of acid rain.

Title VI of the CAA Amendments of 1990 seeks to reduce threats to the stratospheric ozone layer by phasing out production and use of chlorofluorocarbons (CFCs), halons, and other widely used chemicals believed to contribute to global warming.

Clean Water Act (CWA)

The Clean Water Act (CWA) is a Federal statute that addresses the quality needs of the nation's waterbodies, with regard to both human and environmental concerns. It is written to control known, possible, and unknown toxins and pollutants through proper use and disposal. The objective of the CWA is to restore and maintain the chemical, physical, and biological integrity of the nation's water.

Appendix B-65, Toxic Pollutants, is one of the earliest written and most important tables for identifying toxic pollutants. These toxic pollutants are composed of organics, inorganics, heavy metals, cyanogens, halogens, and PCB compounds. The amended list is found in 53 FR 46015, October 17, 1988, and is incorporated by reference in CWA section 307(a)(1).

In addition, the CWA designates under section 301(2)(C)(f) categories of pollutant emissions for organic chemicals, plastics, synthetic fibers, and pesticides. Besides listing toxic pollutants, the CWA lists 27 categories of sources that discharge toxic waste, e.g., pulp and paper mills, dairy product processing, textile mills, feedlots, electroplating industries, plastic and synthetic materials manufacturing, and petroleum refining.

The term TTO, "total toxic organics," is used to describe a group of chemicals with quantifiable amount greater than 0.01 milligram per liter. This expanded list of 126 toxic organics was set up to establish effluent limitations under CWA section 301(b)(2)(C).

CWA subpart D, section 414.40 has a list of Standard Industrial Code (SIC) 28213 thermoplastic resins and thermoplastic resin groups that are applicable to the process wastewater discharges resulting from the manufacture of thermoplastic resins.

CWA subpart G, section 414.70 lists bulk organic chemicals that are associated with wastewater resulting from the manufacture of SIC 2865 and 2869 bulk organic chemicals and organic chemical groups. The list is broken down into five groups: aliphatic, amine and amide, aromatic, halogenated, and other organic chemicals.

Subpart F, section 414.60, Commodity Organic Chemicals lists the applicable compounds in process wastewater resulting from the manufacture of SIC 2865 and 2869 commodity organic chemical groups.

Comprehensive Environmental Response, Compensation and Liability Act (CERCLA)

The Comprehensive Environmental Response, Compensation and Liability Act (CERCLA) is a national law designed to regulate the cleaning up of environmental contaminations made before and not covered under the Resource Conservation and Recovery Act (RCRA). CERCLA has been termed "Superfund" because of the large amounts of money appropriated by Congress to clean up the nation's environment. As an amendment to CERCLA, the Superfund Amendments and Reauthorization Act (SARA) gives CERCLA new strength by providing new cleanup standards, schedules, and provisions aimed at federal facilities and increased settlement, liability, and enforcement powers for the EPA and citizens.

In 40 CFR 302.4 is a list of hazardous substances and reportable quantities that are the chemical compounds regulated by CERCLA; the Chemical Abstract Services Reference Number (CASRN) of each compound; its synonym; the statutory source for designation of the hazardous substance under CERCLA-CWA, CAA, and RCRA; and the reportable quantities. Each hazardous substance is listed in alphabetical order and grouped with its respective family, i.e., antimony and compounds, arsenic and compounds, polychlorinated biphenyls (PCBs), spent halogenated solvents used in degreasing, wastewater treatment sludges from electroplating operations, and spent catalyst from the hydrochlorinator reactor in production of 1,1,1-trichloroethane.

CERCLA also regulates radionuclides, and in Appendix B of 40 CFR 302.4 is a table with an alphabetic listing of the hazardous substances along with their reportable quantities specified in curies.

Also in 40 CFR 302.4, Appendix B of CERCLA is the "List of Extremely Hazardous Substances and Their Threshold Planning Quantities." This table is listed according to CASRN and has a column for notes that can include such information as Threshold Planning Quantity (TPQ), the statutory reportable quantity, and toxicity criteria.

Resource Conservation and Recovery Act (RCRA)

The Resource Conservation and Recovery Act (RCRA) was enacted to fill the regulatory pollution control gap between the Clean Air Act and the Clean Water Act. Despite its name, RCRA's primary purpose has been to control the disposal of hazardous and solid wastes generated by various manufacturing processes and the air and water pollution control devices installed for those processes. Waste management categories within RCRA include the following subtitles: Subtitle C—Hazardous, Subtitle D—Nonhazardous, and Subtitle J—Medical

Waste. All hazardous wastes must first meet the definition of a solid waste. A solid waste is any garbage, refuse, sludge, and other discarded material, including solid, liquid, semi-solid, or contained gaseous material resulting from industrial, commercial, mining, and agricultural operations, and from community activities, except domestic sewage, irrigation return flows, or industrial discharges controlled as point sources under the Federal Water Pollution Control Act or materials controlled under the Atomic Energy Act. Discarded solid wastes include abandoned materials, recycled materials, and inherently waste-like (dioxin-containing) materials.

A hazardous waste is defined as any solid waste, or combination of solid wastes which, because of its quantity; concentration; or physical, chemical, or infectious characteristics, may (1) cause, or significantly contribute to an increase in mortality or an increase in serious irreversible, or incapacitating reversible, illness; or (2) pose a substantial present or potential hazard to human health or the environment when improperly treated, stored, transported, or disposed of, or otherwise managed. Hazardous wastes include those that either exhibit a hazardous waste characteristics, as defined in subpart C of 40 CFR Part 261 or are a hazardous waste listed in subpart D.

Characteristic hazardous wastes, as defined in subpart C of 40 CFR Part 261, include those which exhibit one of the following characteristics:

- Ignitability, as defined in section 261.21, applies to solid wastes that are capable of causing fires during routine handling and/or significantly increasing the dangers of a fire once one is started.
- Corrosivity, as defined in section 261.22, applies to liquid wastes with a pH of less than 2 or more than 12.5 and solid wastes with the ability to corrode steel.
- Reactivity, as defined in section 261.23, applies to the capability of a waste to explode, undergo violent chemical change in a variety of situations, or react violently with water to produce toxic fumes or vapors.
- Toxicity, as defined in section 261.24, applies to the capability of a solid waste to release into water any of 40 toxic constituents in concentrations above regulatory levels established by the EPA. The standard test method for this characteristic is known as the Toxicity Characteristic Leaching Procedure (TCLP).

Specific wastes have been identified as hazardous by the EPA because of known hazardous characteristics. These types of wastes and the locations of their lists in 40 CFR are given below:

- Manufacturing wastes from nonspecific sources (F code wastes) are listed in section 261.31.
- Manufacturing wastes from specific industrial processes (K code wastes) are listed in section 261.32.
- Discarded chemical products or intermediates that are acutely toxic wastes (i.e., if the LD_{50} is less than 50 mg/kg) (P code wastes) are listed in section 261.33(e).
- Discarded chemical products or intermediates that present risks of chronic toxicity from exposure (U code wastes) are listed in section 261.33(f).

TOXIC SUBSTANCES CONTROL ACT (TSCA)

The Toxic Substances Control Act (TSCA) includes regulations and testing requirements for every chemical substance that is manufactured for commercial purposes in the United States or imported for commercial purposes. The following chemicals are regulated under TSCA with respect to processing, use, and disposal, as well as warnings and instructions that must accompany the substance when distributed: polychlorinated biphenyls (PCBs), fully halogenated chlorofluoroalkanes (CFCs), and asbestos.

TSCA Part 761 establishes prohibitions of, and requirements for, the manufacture, processing, distribution in commerce, use, disposal, storage, and markings of PCBs and PCB items. Substances with PCBs that are regulated by this rule include, but are not limited to, dielectric fluids, contaminated solvents, oils, waste oils, heat transfer fluids, hydraulic fluids, paints, sludges, slurries, dredge spoils, soils, materials contaminated as a result of spills, and other chemical substances or combinations of substances.

TSCA Part 746 prohibits the manufacture, processing, and distribution of CFCs as aerosol propellants, except for export. Two other classes of exemptions for CFC propellants are (1) for use in an article which is a food, food additive, drug, cosmetic, or exempted device, and (2) for essential and exempted uses listed in sections 762.58 and 762.59.

TSCA Part 763, subpart D requires reporting by persons who manufacture, import, or process asbestos. Part 763, subpart I prohibits the manufacture, importation, processing, and distribution in commerce of the asbestos-containing products identified and at the dates indicated. This subpart requires that products subject to this rule's bans, but not yet subject to a ban on distribution in commerce, be labeled.

One of the major goals of TSCA is to develop test data which are necessary to determine whether chemical substances and mixtures present an unreasonable risk to health or the environment. Under Section 4 of TSCA, the EPA can require chemical manufacturers, importers, and processors to conduct and pay for those tests. TSCA Part 799 identifies the chemical substances, mixtures, and categories of substances and mixtures for which data are to be developed, specifies the persons required to test, specifies the test substance(s) in each case, prescribes the tests that are required, and provides deadlines for submission of reports and data to EPA. Part 766 identifies testing requirements to ascertain whether certain specified chemical substances may be contaminated with halogenated dibenzodioxins (HDDs)/dibenzofurans (HFDs), as well as requirements for reporting these analyses.